Ethical Issues of Information Systems

Table of Contents

Foreword

Since the beginning of time, new technologies have evolved to influence society, creating new possibilities and making our lives easier. Information Technology (IT) is one of these. IT has created new social and ethical dilemmas by influencing and producing situations that can conflict with existing laws, rules, traditional ethical and moral principles as well as cultural norms and values.

Our daily lives are dependent on information technology. The same technology that can be used to enhance our lives in turn may make our lives very difficult. As IT technology rapidly advances, it could create significant social problems. While the human intellect and societal order is creeping onward at linear speed, technology is racing exponentially. The rate of technology growth and the doubling speed of computers every 18 months makes it very difficult for humans to adjust to change. Recently, the famed British physicist *Stephen Hawk* said, "if humans hope to compete with the rising tide of artificial intelligence, they will have to improve through genetic engineering." He added, "So the danger is real that they (computers) develop intelligence and take over the world." This statement shows the significance of either controlling the technology or changing the human behavior toward it.

Today, we see the negative effect of IT with activities such as invasion of privacy, infringement of property rights, breach of security, and the denial of access to the rightful owners of the information in our lives. If this trend continues, the advancement of computing technology will add to the frustrations of IT users.

Historically, IT professionals as a group have not been overly concerned with questions of social misuses related to computing. As a matter of fact, a majority of negative affects of IT related activities are ignored by the IT community, businesses, and the industry either as a consequence of publicity or because they are complacent about computers. Information technology, especially the Internet, was built on the assumption of an open society and trust. The assumption was that scientists trust each other and can share their discovery with the other members of the society free from negative consequences such as violations of security.

If we assert that every IT professional has certain rights, we should also assert that the same group of people also has responsibilities, which include responsibilities to oneself, employer, profession, and the community at large. Individuals should have the right to privacy, accuracy, property, and accessibility to information systems.

While individuals and professional IT organizations have tried to address these issues, the whole IT community remains on the sidelines and needs to address social problems arising from IT and develop a set of guidelines. The IT community deserves to have social integrity and autonomy. Integrity is defined as "a matter of self-respect," such as being true to oneself and living responsibly in light of one's limitations. Self-respect provides society the opportunities to develop the best social and professional integrity and allow greater personal integrity on the part of its members.

To protect society from the infringement of its rights by certain elements, a set of rules and laws must be written. But in this global environment, legislating the Internet has its own drawbacks. First of all, which country's laws are applicable when the client and the server machines are located in two different countries thousands of miles apart? And since what is illegal in one country may not be illegal in another country, which entity will judge the lawbreakers? How can laws be unified in an environment that must deal with multiple languages, multiple cultures, multiple governments, and multiple geographical entities? How will the difference between people with different interests, customs, mores, and traditions be resolved? Who is going to speak on behalf of the community as a whole?

As Deborah Johnson and Helen Nissenbaum indicated, "We must decide these issues of ethics as a community of professionals and then present them to society as a whole. No matter what laws are passed, and no matter how good security measures might become, they will not be enough for us to have completely secure systems." The members of the IT community need to understand the importance of respecting privacy and data ownership.

The best way to implement and enforce ethical and social responsibilities in the IT environment is to educate IT professionals. Ethical and societal issues in information technology education should produce IT graduates aware of social and ethical issues created by computers. The IT professionals need to understand their role in contributing to society and human well-being is an essential aim of computing professionals. Also, the minimization of negative consequences of computing systems, including threats to health and safety, should be the paramount goal of all IT professionals.

Ali Salehnia
Brookings, South Dakota, U.S.A.
September 17, 2001

Preface

As information systems use becomes more widespread and more individuals and organizations rely on the Internet as a means of conducting business, it becomes ever more important to assure that the Internet is a place where privacy is protected. Additionally, as organizations rely more on information systems, they become vulnerable to attacks on these precious technologies. These are just some of the ethical issues professionals face when dealing with information systems and emerging technologies. Although often overlooked, ethical decision-making is an important issue for all organizations and individuals in the arena of information technologies. In order to better understand the ethical dilemmas facing professionals and private citizens and appreciate their consequences, researchers, practitioners and academics must have access to the latest thinking and practice concerning ethics and information systems. The following chapters contain the most recent research and practices in defining and applying ethical standards. They cover the spectrum of thinking from applying Kierkegaard's philosophy to Internet education to appreciating the factors that lead to software piracy. There is something that will benefit everyone concerned with the ethics of information systems.

Chapter 1 entitled, "Internet Privacy: Interpreting Key Issues" by Gurpreet Dhillon and Trevor Moores of the University of Nevada, Las Vegas (USA) systematically identifies major Internet privacy concerns. The chapter's primary purpose is to identify issues related to maximizing Internet privacy. The authors present various researchers' definitions of the concept and then proceed to discuss a study they conducted attempting to further understand this concept.

Chapter 2 entitled, "One Size Does Not Fit All: Potential Diseconomies in Global Information Systems" by Gerald Grant of Carleton University (Canada) explains how global information systems may lead to diseconomies of scale. The author discusses the differences in companies as well as the internal and external factors that influence an organization's technology needs and ability to implement information systems. The author proposes careful scrutiny of costs and associated paradigm changes that a company will undergo to implement new technologies.

Chapter 3 entitled, "Some Internet and E-Commerce Legal Perspectives Impacting the End User" by Peter Mykytyn of Southern Illinois University (USA) discusses two legal issues that can confront today's end users as they do business over the Web. The author addresses the important issues of contract law and jurisdictional issues. These important issues are becoming ever more significant because the current laws were enacted in a world where goods and services were the primary commodities of business, not information.

Chapter 4 entitled, "A New Approach to Evaluating Business Ethics: An Artificial Neural Networks Application" by Mo Adam Mahmood, Gary Sullivan and Ray-Lin Tung of the University of Texas, El Paso (USA) presents a new approach to classifying, categorizing, and analyzing ethical decision situations. The chapter offers a comparative analysis of artificial neural networks, multiple discriminate analysis and chance. This analysis shows that artificial neural networks predict better in both training and testing phases and offer a promising alternative to traditional analytical tools.

Chapter 5 entitled, "Copyright, Piracy, Privacy and Security Issues: Acceptable or Unacceptable Actions for End Users?" by Jennifer Kreie and Timothy Paul Cronan of the University of Arkansas (USA) examines factors that may influence decision-making based upon models of ethical behavior. The results of an empirical study conducted to determine end user perceptions of acceptable or unacceptable behavior and which factors influenced a person's judgment of the acceptability of a behavior are reported in this chapter. The results reported indicate that the factors that influenced perceptions were based upon the characteristics of the ethical dilemma.

Chapter 6 entitled, "Ten Lessons that Internet Auction Markets Can Learn from Securities Market Automation" by J. Christopher Westland of Hong Kong University of Science and Technology (Republic of China) explores the automation of three emerging market exchanges: The Commercial Exchange of Santiago, The Moscow Central Stock Exchange and Shanghai's Stock Exchange and compares Internet models of retailing with the older proprietary networked markets for financial securities.

Chapter 7 entitled, "The Societal Impact of the World Wide Web—Key Challenges for the 21st Century" by Janice Burn of Edith Cowan University (Australia) and Karen Loch of Georgia State University (USA) documents the current state of information technology diffusion and connectivity and the related factors such a population density, cultural attitudes and gross domestic product. The chapter then looks specifically at who the "haves" and the "have-nots" with regards to technology are. Finally, the authors then offer concrete suggestions about how the Internet may be used to bridge the gap between the advantaged and the disadvantaged.

Chapter 8 entitled, "Method Over Mayhem in Managing e-Commerce Risk" by Dieter Fink of Edith Cowan University (Australia) identifies the differences between risk management approaches for older information technology systems and those required for e-commerce. The chapter discusses the benefits and critical success factors for an e-commerce risk management methodology. The authors then recommend a program of research to make risk management more dynamic and interactive particularly for the operational aspects of e-commerce.

Chapter 9 entitled, "Why Do We Do It If We Know It's Wrong? A Structural Model of Software Piracy" by Darryl Seale of the University of Nevada, Las Vegas (USA) examines the predictors of software piracy, a practice estimated to cost the software industry nearly $11 billion in lost revenue annually. The chapter develops a structural model which suggest that social norms, expertise required, gender and computer usage have direct effects on self-reported piracy. The author discusses the theoretical and practical implications for the design and marketing of software.

Chapter 10 entitled, "Ethical Issues in Software Engineering Revisited" by Ali Salehnia of South Dakota State University and Hassan Pournaghshband of Southern Polytechnic State University (USA) looks at each step in the software engineering process and how these steps affect the reliability and safety of the analysis, design and implementation of software. The authors then examine the ethical aspects of software and systems development.

Chapter 11 entitled, "Cyberspace Ethics and Information Warfare" by Matthew Warren of Deakin University and William Hutchinson of Edith Cowan University (Australia) examines the evolution of information warfare from a group of young individuals, hackers to organized individuals, corporations, government agencies, organized crime and terrorists wrecking havoc in the information age. The chapter looks at specific tactics of information warfare and future trends of these attacks.

Chapter 12 entitled, "A Conversion Regarding Ethics In Information Systems Educational Research" by Mark Campbell Williams of Edith Cowan University (Australia) reflects on the author's own heuristic and psychologically-oriented self study concerning some ethical improprieties committed during the data collection phase of an information systems educational research program. The chapter reports a self-dialogue concerning the issues of ethics and research and asserts that ethical paradigms are especially important when investigating a new media where acceptable ethical practices have not yet been established.

Chapter 13 entitled, "Software Piracy: Are Robin Hood and Responsibility Denial at Work?" by Susan Harrington of Georgia College and State University (USA) examines the factors that lead to the unethical use of computers. Guided by existing ethical decision-making models, this chapter examines the individual characters that are underlying causes for persistent abuse. The author reports the results of a study which specifically examined the characteristics of responsibility denial and the Robin Hood Syndrome.

Chapter 14 entitled, "Social Issues in Electronic Commerce: Implications for Policy Makers" by Anastasia Papazafeiropoulou and Athanasia Pouloudi of Brunel University (United Kingdom) examines social issues related to electronic commerce policy-making and presents two fundamental social concerns related to policy-making: trust and digital democracy. The chapter then discusses these concerns as they relate to the policy issues concerning network technologies and presents the implications of these concerns for policy-making in electronic commerce.

Chapter 15 entitled, "Kierkegaard and the Internet: The Role and Formation of Community in Education" by Andrew Ward of the University of Minnesota and Brian Prosser of Fordham University (USA) applies Kierkegaard's philosophy to Internet technologies. The author looks at the effect of communication via the Internet and analyzes the emergence of virtual classrooms in the contest of Kierkegaard's philosophies.

Chapter 16 entitled, "Manufacturing Social Responsibility Benchmarks in the Competitive Intelligence Age" by James Orton of the University of Nevada, Las Vegas (USA) explores the emergence of social responsibility benchmarks in the competitive intelligence age. The chapter reports on attempts by competitive intelligence agents in the United States and France to manufacture social responsibility benchmarks in the context of covert operations, competitive strategy, corporate intelligence, economic security, economic intelligence and economic warfare.

Chapter 17 entitled, "Strategic and Ethical Issues in Outsourcing Information Technologies" by Randall Reid of the University of Alabama and Mario Pascalev of the Bank of America (USA) identifies major ethical problems and proposes guidelines for ethical conduct in the process of outsourcing information technology. The chapter discusses the benefits and models of outsourcing information technology and looks at ethical literature in general and professional organizations' codes of ethics in particular. The authors then present and analyze a case study involving IT outsourcing. Finally, the authors suggest general ethical guidelines for outsourcing models.

Chapter 18 entitled, "Ethics, Authenticity and Emancipation in Information Systems Development" by Stephen Probert of Cranfield University (United Kingdom) describes research on the philosophical concept of authenticity used as a framing device for providing an interpretation of ethical and practical action by information systems professionals. The chapter discusses the implications of an IS professional choosing to do research in an authentic manner rather than doing so in adherence with a code of professional ethics or a series of methodological precepts.

Chapter 19 entitled, "On the Role of Human Morality in Information Systems Security: From the Problems of Descriptivism to Non-Descriptive Foundations" by Mikko Siponen of the University of Oulu (Finland) discusses various ethical frameworks and explores objections to the use of ethics as a means of protection based on cultural relativism. The chapter then offers an alternative approach based on a theory of non-descriptivism and discusses the implications of this alternative.

Chapter 20 entitled, "The Government 'Downunder' Attempts to Censor the Net" by Geoffrey Sandy of Victoria University (Australia) reports the results of an analysis of the primary sources of an Australian bill, the Broadcasting Services Amendment. The purpose of this bill is to regulate access to content that is offensive to a reasonable adult and unsuitable for children. Specifically, the chapter discusses the important issues that were addressed in the parliamentary hearings and debates. The chapter also comments on the success of the legislation after eight months of operation.

Chapter 21 entitled, "The Genetic Revolution: Ethical Implications for the 21st Century" by Atefeh McCampbell of Florida Institute of Technology and Linda Moorhead Clare of Information Technology Group (USA) defines the practice of DNA analysis and identifies the ethical considerations of human genetic testing in the workplace. The discussion in the chapter is based on a survey conducted to determine the view and level of knowledge among business processionals in the workplace on the ethical considerations of genetic testing.

The ethical dimension of information systems encompasses all facets of information technology: research, practice and development. The chapters in this timely new book present theoretical frameworks for developing ethical standards and applying them to information technology research and data analysis. They further discuss practical applications of these frameworks and offer concrete suggestions about how to incorporate ethical thinking into all aspects of organizational life. From legislation censoring the Internet to the threat of information warfare, these chapters are relevant and timely in their understanding and application of the new ethical challenges in the information era.

IRM Press
January 2002

Chapter 1

Internet Privacy: Interpreting Key Issues

Gurpreet S. Dhillon and Trevor T. Moores
University of Nevada, Las Vegas, USA

The phenomenal growth in Internet commerce in recent years has brought privacy concerns to the fore. Although privacy as a concept has been well understood with respect to brick and motor businesses, there is limited research in identifying major issues of concern related to Internet privacy. This paper systematically identifies the major Internet privacy concerns. Data for the study was collected through two panels and subjective evaluation.

INTRODUCTION

The Internet has transformed the way in which goods are bought and sold. Forrester Research predicts retail sales on the Internet to grow from less than 1 percent in 1999 to 6 percent by 2003. According to Gartner Group, convenience and time saved are two of the main incentives for users to buy online. At the same time, however, research conducted by Price Waterhouse Coopers suggests that during the 1999 Christmas season, 18% of all customers who purchased online were 'dissatisfied' with their experience. A Business Week/Harris Poll (see *Business Week* of March 20, 2000) survey reported that 41% of online shoppers were very concerned over the use of personal information. Among the people who go online but have not shopped, 63% were very concerned. Clearly, as Keeney (1999) suggests, maximizing privacy is a fundamental objective related to Internet commerce.

The purpose of this paper is to identify issues related to maximizing Internet privacy. The paper is organized into five sections. Following a brief introduction, section two explores the notion of Internet privacy and how various researchers have attempted to understand the concept. Section three presents the study design.

Section four is a discussion of research findings. Section five presents the conclusions.

INTERNET PRIVACY

Internet privacy can be defined as the seclusion and freedom from unauthorized intrusion. The key word in the definition is 'unauthorized'. Although we may not like that our personal information regarding our purchases and habits to be monitored and stored in databases around the country, we are at least usually aware that it's happening. However an unauthorized intrusion to collect personal data marks the beginning of privacy infringement. Various opinion polls have shown increasing levels of privacy concerns (Equifax, 1990, 1992). The 1992 Equifax study reports a survey indicating nearly 79% of the Americans being concerned about personal privacy and 55% suggesting that security of personal information was bound to get worse by year 2000. Indeed this has happened. Fairweather & Rogerson (2000) report that it is technically easier than ever before to gather and search vast amounts of personal data. Hence it has become easy to track individuals across the globe as they leave the data shadow behind – through the use of gas stations, cash machines, logging on to check email.

A March 1999 Federal Trade Commission (FTC) survey of 361 Web sites revealed that 92.8% of the sites were collecting at least one type of identifying information, such as an address. Furthermore 56.8% of the sites were collecting at least one type of demographic information. The FTC study also found that over one third of the sites did not have a privacy disclosure notice on the site. Even in cases where the privacy disclosure notice had been posted, only 13.6% were following the FTC's fair information practice guidelines.

Previous literature on privacy - not necessarily Internet privacy–has critiqued the majority of opinion surveys based on the assumption that information privacy is not a uni-dimensional construct, i.e., focusing on the level of concern alone, rather than understanding the nature of concern. In response, Smith et al. (1996) suggest four dimensions of the construct "individuals' concerns about organizational practices in managing information privacy." These factors were: collection, unauthorized secondary use, improper access, and errors. Smith et al.'s (1996) research, although providing a very useful instrument to measure individuals' concern about information privacy, does not necessarily consider privacy issues in relation to Internet use. Clearly the use of the Internet to conduct business has gained prominence in recent years and the converging trends, competitive and technological, pose interesting privacy challenges (cf. Culnan & Armstrong, 1999).

There are two reasons for an increased importance of Internet privacy concerns, as opposed to simple information privacy issues relevant to any brick-and-mortar business. First, the increasingly competitive business environment is forcing companies to collect a vast amount of personal information. Many a time there is good intent in doing so, since many businesses may seriously want to customize their products and services for the benefit of the consumer. However the security of personal data and subsequent misuse or wrongful use without prior permission of an individual raise privacy concerns and often end up in questioning the intent behind collecting private information in the first place. Second, the advances in information technology have not only made it possible to record personal information at the point of sale, but also map the patterns of online behavior. Although this is a useful marketing ploy (Bessen, 1993; Glazer, 1991), it certainly overwhelms the customer and hence there are numerous privacy concerns. Similar issues about overwhelming the customer through excessive use of technology have been voiced in the literature (see Dhillon & Hackney, 1999; Ciborra, 1994).

With respect to the two reasons identified above, the question of fairness in collecting personal information needs to be understood adequately. Fairness, with respect to Internet commerce, can be considered at two levels. As Glazer (1991), and Milne & Gordon (1993) contend, fairness could either be a component in the 'social contract' or related to the procedure followed for a particular activity (Lind & Tyler, 1988; Folger & Bies, 1989). When individuals willingly disclose personal information for non-monetary gains, such as higher quality service, privacy concerns are limited as long as the concerned organization upholds its side of the social contract. Individuals will clearly continue engaging in the social contract as long as the benefits exceed the risks, to a point where an individual begins trusting the organization. This is evidenced by many of the new generation Internet businesses. Barnesandnoble.com and Yahoo, for example, have clear-cut privacy policies, thereby facilitating in developing trust over a period of time. On the other hand ediets.com believes in overwhelming the customer with emails and offers once personal details have been recorded.

Fairness is also linked to the procedure that might be followed in a particular activity. Clearly fairness of the procedure, as opposed to the nature of the outcome (Lind & Tyler, 1988), is a clear determinant of the level of privacy concern an individual might have. Some Internet businesses are now beginning to place importance on procedural fairness. In many cases the Web sites first give a notice as to why personal information is being collected, its usefulness and the manner in which it would be kept secure, then the consent is sought as to the manner in which an individual's personal information would be used. As would be evident, procedural fairness is closely coupled with social contract and trust. If an individual feels that in spite of procedural fairness, the social contract in the exchange of private information is not maintained, it would clearly lead to loss of trust and integrity of the organization. On the other hand if an individual willingly gives private information in lieu of some social or economic benefit, but the procedure used in collecting and maintaining the information is not fair, again it would lead to concerns about privacy infringement, trust and integrity of the process.

Given an understanding of various aspects of Internet privacy, as discussed in the literature, our intention is to understand the various issues that could be of potential concern for individuals. The next section describes the multi-method adopted to identify such issues.

STUDY DESIGN

In identifying issues related to individuals' concerns about Internet privacy, we set out to use a combination of two methodological approaches. The first relates to steps 1 and 2 as described by Schmidt (1997) while the second is related to the identification of means and fundamental objectives as described by Keeney (1999). A combination of these two approaches helped us to generate a list of issues that are of significant concern for individuals with respect to Internet privacy. Further research would enable us to validate the preliminary list and develop an instrument that would be useful in assessing the level of Internet privacy concern for an individual with respect to a particular online business.

This study was designed to span two main phases. Phase one followed Schmidt's (1997) approach to (a) discover relevant issues and (b) determine the most important issues. Phase two of the study followed Keeney (1999) in identifying the fundamental Internet privacy objectives of individuals and means objectives in achieving the fundamental Internet privacy objectives. Essentially Keeney's concepts were used to classify the output of Schimidt's second step.

Keeney (1999) stresses the importance of defining a decision context when identifying the objectives. He contends that the fundamental objectives together with the decision context provide a decision frame. Furthermore a decision context defines the alternatives to consider for a specific decision situation. In our study the decision context was the maintenance of individual privacy with respect to Internet use. We defined our overall objective as maximizing Internet privacy for individuals. According to Keeney, the decision context would imbed in itself a number of means objectives. These would be objectives that individuals would have with respect to maximizing Internet privacy. Our task was to not only identify fundamental Internet privacy objectives, but also all possible means objectives. We also wanted to rank all objectives in order of importance.

Phase 1

Our first step in Phase 1 was to unearth as many issues as we could from a panel of experts. Panel members, eleven in total, were invited to a brainstorming session and were asked to identify all possible objectives they would have in maximizing Internet privacy. The actual elicitation of objectives followed a 40-minute general discussion on Internet privacy issues and was moderated by the first author. The panel was representative of various experts in the field. There was one attorney, one former policeman, one network administrator, one sales and marketing professional, one software engineer, one dot-com entrepreneur, two full-time students and three ardent Internet users who had considerable experience in purchasing online.

The brainstorming session lasted little over an hour and the panelists identified 144 concerns/objectives. At this stage it was hard to differentiate whether these were merely concerns or were in fact Internet privacy objectives. Following the data collection exercise, the authors consolidating the list of objectives and posted it on their web site. The consolidation process produced 70 objectives. The panelists were invited to visit the Web site to refine, add to or suggest deletions from this list. An online bulletin board was used to capture the responses. The respondents added another 15 objectives to the consolidated list to produce a total of 85 objectives.

Our second step was to determine the most important issues. We presented our list of 85 objectives to a group of 16 IS executives. These executives represented five different industries: government, hotel, pharmaceutical, health care and IS consulting. The average work experiences of the IS executives was five years. The group was asked to rank the top 10 issues from the list of 85 objectives. We followed the guidelines of selecting at least 10% following Schmidt (1997). No ties were allowed. The results were consolidated and presented once again to the group. Open discussion resulted in clearly identifying the top five objectives. These objectives appear in Table 1 in rank order.

Table 1: The top five Internet privacy issues

Rank #	Issue stated as objective
1	Companies should not sell personal information.
2	Adequate measures should be in place to prevent theft of personal information by a third party.
3	Eliminate the chance of 'losing' personal files.
4	Maximize security to deter 'hackers' from destroying the data.
5	Eliminate spam.

Table 2: Internet privacy means objectives

Increased awareness to have firewall protection	**Boycott companies who do not have a privacy policy**
Provide credit card security assurance by third parties	**Use encryption in email communications**
Provide guarantees from shopping sites	**Online businesses should not collect personal information**
Enact stronger laws to protect consumer privacy	**Watch children online**
Tougher laws to protect consumer ID theft	**Strict penalties for violators of personal privacy**
Facilitate self-policing the Internet community	**Establish international standards on privacy**
Prosecute violators of laws	**Check authenticity of an online business prior to purchase**
Make spam illegal	**Increase self determination in providing personal information online**
Businesses should be required to have a privacy policy	**Providing personal information online should be discretionary**

Phase 2

Phase 2 involved the identification of means objectives with respect to maintaining Internet privacy. The argument used in identifying the means objectives was that any objective that was not a top issue of concern clearly contributed in some way to one of the top issues. The remaining 80 issues from Phase 1 were subjectively evaluated with an intent to formulate means objectives. This was a two-step process. First, issues with a similar meaning were clustered together. Second, a judgment was made whether an issue had merit to be a means objective on its own or was merely a part of a larger means objective. This process resulted in 18 means objectives. The majority of the remaining 62 issues were condensed into one of the 18 means objectives. Others did not necessarily relate to our overall objective and were hence not included. The means objectives (in no particular order) appear in Table 2.

DISCUSSION OF RESEARCH FINDINGS

This section presents a discussion of key Internet privacy issues. The intent is to provide an explanation of the top issues identified by respondents in this study in light of the literature and the means objectives.

The top issue is the potential for a Web site to sell the details of online consumers to a third party. While the use of personal information to further the cause of businesses has become a competitive necessity, the issue raised here suggests that the burden resides with online businesses to ensure that confidentiality of personal information collected is maintained. As identified in Section two of this paper, individuals may be willing to provide their personal information as long as they are receiving some benefit, i.e., increased customer service. This means that online businesses need to ensure 'procedural fairness' (Culnan & Armstrong, 1999). Individuals may not be interested in giving out personal information on first contact with an online business, but a trust may develop over a period of time. Hence, as identified in this research (see table 2), it is important for businesses to self-determine what they should divulge to a third-party. Moreover providing personal information online should be discretionary. Such actions would go a long way to enhance the credibility and integrity of online businesses and enable them to remain competitive ethically.

Establishing adequate measures to prevent identification theft is another critical concern identified by this research. This issue can be addressed at two levels: 1) the security of personal information once it has been collected; and, 2) establishing tougher laws to prevent consumer ID theft. When dealing with security of information internally, establishment of a security policy and a general culture of trust and high integrity will be beneficial (Dhillon & Backhouse, 2000), as well as other organizational issues that are beyond the scope of this paper (see Dhillon, 1997 for details). There is also a need to have tougher Internet privacy laws, such that violators could be adequately prosecuted. At the present time, there is no doubt that US and European governments are responsive to Internet privacy demands. Research has shown that increased levels of privacy regulation are a function of the level of data processing environment in a particular country and increased government involvement in privacy protection (see Milberg et al., 2000; Flaherty, 1989).

Several of the fundamental issues identified by the respondents in this research related to establishing adequate measures to protect information from inappropriate sale (issue #1), but also from accidental loss (issue #3), and from deliberate attack by 'hackers' (issue #4). In the US, the FTC has also given due credence to the protection of personal financial information. The onus however has been placed on individual organizations to ensure responsible data manipulation, implementing encryption standards and maintaining secure servers. Although many online businesses, especially banks, have security very high on their agenda, many other businesses with an aspiration to develop customer confidence have turned to third parties for 'seals of approval'. In recent years, one means has been the Web Assurance Seals. Organizations such as the Better Business Bureaus, Certified Public Accounting Firms, and/or organizations devoted to security, privacy or dispute resolution award seals of assurance to Web sites that meet certain criteria. The seals of assurance cover such areas as privacy, security, transaction integrity/completeness, business disclosures, quality control processes, and consumer recourse. These seals often highlight the close relationship between privacy and security issues. Clearly there were concerns with maintaining privacy of personal information when connected through a cable modem, from viruses and Trojan horses. Whether deliberate or accidental, the issue raised here is that any data provided to a Web site must be protected.

An issue related with electronic communication is that of spam. Receiving unsolicited email is certainly an invasion of privacy. Some businesses feel that inundating consumers about their products and services is going to increase their sales. However the converse may be true. Unsolicited emails are irritating and our respondents seemed to have a very strong opinion about them. Some of the popular internet e-mail services such as Hotmail and Yahoo! now include a "spam guard" that detects bulk mail and automatically directs the spam to the trash folder. There are now also facilities to block mail from the more technically-savvy spammers that do not use bulk mail addresses in their e-mail header. According to the respondents in this study, spam e-mail is seen as a sufficiently irritating phenomenon that there are calls to make spam illegal.

In this section we have discussed various aspects of Internet privacy which our respondents considered to be fundamental concerns. Clearly there are no simple answers to the issue of Internet privacy, but our fundamental issues would help in starting a dialogue.

CONCLUSION

This paper has identified an individual's concerns with respect to Internet privacy. Five fundamental and eighteen means objectives were identified, essentially suggesting that in order to adequately manage the fundamental concerns, concrete steps have to be taken with respect to the means objectives. While the governments and organizations gear up their resources to tackle Internet privacy concerns, it is prudent to engage in self-regulation. A way ahead could be through the creation of private rights of action for individuals who have been harmed. Following on from research presented in this paper, work is needed to develop measures to assess the extent to which individuals are comfortable with Internet privacy within the context of a particular business. This would help businesses to create new policies and reassess existing ones.

REFERENCES

Bessen, J. (1993) Riding the marketing information wave, *Harvard Business Review,* 71, 5 (September-October), 150-160.

Ciborra, C. (1994) Market support systems: theory and practice, in G. Pogorel (ed.), *Global telecommunications strategies and technological changes,* North-Holland, Amsterdam, 97-110.

Culnan, M. J., and Armstrong, P. K. (1999) Information privacy concerns, procedural fairness, and impersonal trust: an empirical investigation, *Organization Science,* 10, 1, 104-115.

Dhillon, G. (1997) *Managing information system security*, Macmillan, London.

Dhillon, G., and Backhouse, J. (2000) Information system security management in the new millennium, *Communications of the ACM,* 43, 7, 125-128.

Dhillon, G., and Hackney, R. (1999) IS/IT market support systems: augmenting UK primary care groups, *Topics in Health Information Management,* 20, 2.

Equifax. (1990) *The Equifax report on consumers in the information age,* Atlanta, GA: Equifax Inc.

Equifax. (1992) *Harris-Equifax consumer privacy survey*, Atlanta, GA: Equifax Inc.

Fairweather, N. B., and Rogerson, S. (2000) *Social responsibility in the information age*, Leicester: CCSR, De Montfort University, UK.

Flaherty, D. H. (1989) *Protecting privacy in surveillance societies*, University of North Carolina Press, Chapel Hill.

Folger, R., and Bies, R. J. (1989) Managerial responsibilities and procedural justice, *Employee Responsibilities and Rights Journal,* 2, 2, 79-90.

Glazer, R. (1991) Marketing in an information-intensive environment: strategic implications of knowledge as an asset, *Journal of Marketing,* 55, 4, 1-19.

Keeney, R. L. (1999) The value of Internet commerce to the customer, *Management Science,* 45, 4, 533-542.

Lind, E. A., and Tyler, T. R. (1988) *The social psychology of procedural justice*, Plenum Press, New York.

Milberg, S. J., *et al.* (2000) Information privacy: corporate management and national regulation, *Organization Science,* 11, 1, 35-57.

Milne, G. R., and Gordon, M. E. (1993) Direct mail privacy-efficiency trade-offs with an implied social contract framework, *Journal of Public Policy & Marketing,* 12, 2 (Fall), 206-215.

Schmidt, R. C. (1997) Managing Delphi surveys using nonparametric statistical techniques, *Decision Sciences,* 28, 3, 763-774.

Smith, H. J., *et al.* (1996) Information privacy: measuring individuals' concerns about organizational practices, *MIS Quarterly*, June, 167-196.

Chapter 2

One Size Does Not Fit All: Potential Diseconomies in Global Information Systems

Gerald Grant
Carleton University, Canada

Managers, IT practitioners, and IS researchers are easily seduced by the latest information technology wave. Consequently, we tend not to question conventional assumptions about the implementation of IT systems in organizations. Instead of providing managers with directions, IS researchers can sometimes turn into prognosticators of the latest information technology fad. We call on researchers to delve below the surface of new IT trends to expose inconsistencies between technological promises and the reality of deploying information systems in global organizations.

Many IS researchers are turning their attention to the area of global information management (Gallupe and Tan, 1999). This journal is a vehicle for publishing such research work. Interest in integrated global information systems is fueled both by the developments in information and communications technologies and the trends in business towards globalization of products and markets. Conventional wisdom suggests that businesses operating in global markets would benefit from implementing global information systems and achieve economies of scale and scope. This may be true in some cases, but does it hold for all cases? I suggest it may not. In certain cases deploying global IT systems could lead to diseconomies of scale.

A major selling point for implementing global enterprise-wide information systems is the potential for achieving economies of scale and scope in

Previously Published in the *Journal of Global Information Management, vol.8, no.4*, managing the activities of widely dispersed operations. The thrust of most integrated global IT systems deployment efforts is to provide common information firm wide. By deploying common global information systems, firms hope to increase the efficiency of their global business processes and benefit from cross-functional information transfer and sharing. One way of realizing the benefits of common information systems is through the deployment of enterprise-wide systems. Thousands of companies worldwide have implemented enterprise-wide systems. Companies such as Owens-Corning, Colgate Palmolive, Microsoft, and Cisco Systems have reported tangible benefits as a result of deploying an enterprise resource planning (ERP) system. These systems are very complex to deploy and manage and require huge investments to support business process and structural change, IT systems acquisition and deployment, and human resource development.

Business managers typically buy into the concept that implementing common integrated IT systems will lead to efficiency and effectiveness payoffs. They are sold on the idea that simplified processes and better systems will speed data and information flows across the organization. But do organizations always achieve economies of scale by providing common information? One IT manager involved in implementing an ERP system globally suggests that benefits may not be realizable in some circumstances. He suggests that many consultants in their attempt to sell companies on enterprise systems, use stories of ERP implementation success to try to convince managers to spend the many million dollars necessary to implement such systems. In doing so, they may cite examples such as that of Colgate Palmolive where, " before SAP R/3, it took Colgate U.S. anywhere from one to five days to acquire and order, and another one to two days to process an order. Now, order acquisition and processing combined takes four hours, not up to seven days. Distribution planning and picking used to take up to four days; today, it takes 14 hours. In total, the order-to-delivery time has been cut in half" (Kalakota and Robinson, 1999 p. 183).

The implication that any company might be able to achieve similar benefits is seductive. However, this type of reasoning is fraught with difficulties. All companies do not face the same circumstances. Companies are affected differently by both internal and external environmental forces. They have different strategies, structures, business processes, products and culture. A coal mining company with a few large customers may not have to turn around an order in four hours. Should such a company then invest millions of dollars implementing an enterprise information system just to be able to say it can process an order in four hours? Would this be valuable to the customer? Would it make a significant difference to the company if the order was processed in one week instead of four hours?

There are some indications that firms may experience significant diseconomies in attempting to achieve the common information ideal. These diseconomies may arise because of a variety of contingencies presented by firms' external and internal operating environments. Jarvenpaa and Ives (1993, p. 565) identified several contingency factors for managing IT in the multinational firm. These include: (1) the quality of major hardware vendor support; (2) the quality of major software vendor support; (3) the quality of local telecommunications support, (4) pressure for cost savings, (5) senior management support, (6) IT management support, and (7) subsidiary resistance. Other important contingencies may encompass the quality of firm resources (human, technology, financial), as well as the prevailing firm structure and culture, differences in culture and business methods, and quality of global telecommunication networks.

Some senior IT managers are concerned that in many cases organizations have not done enough to determine whether or not deploying common global information systems make economic and operational sense. Davenport (1997) argues that "common information is frequently a high cost strategy, both to implement and to maintain." In one case, a multinational organization decided to implement a global financial information system designed to provide common information across the firm. The idea was to create a "seamless" global company. The firm had a highly decentralized business structure that was itself the result of a major restructuring and rationalization process. This structure required the periodic sharing of summary financial governance and reporting information with corporate headquarters. Business units operated quite autonomously, taking only general direction from corporate headquarters. The decision to implement a common information system globally created a significant misfit between the business operating model and the IT execution model. A centralized IT system was being imposed on a highly decentralized organization. Business unit management resisted the new system because it centralized control at corporate headquarters. At the same time, top managers were reluctant to change the prevailing organizational structure and incentive systems.

The cost of implementation skyrocketed for a number of reasons. One, a common global information system required a supporting computing and telecommunications network. This meant a significant upgrading of computing and networking equipment, facilities, and services. It also meant putting in networking systems where none existed before and where the public telecommunications infrastructure was inadequate. Two, the chosen ERP software required significant modification especially in certain national or regional circumstances. For example the chosen version of the software was not Euro compliant. It had to be extensively modified for operations in Europe. Had the company not adopted the single information system approach, it could have deployed the older version of the software that was Euro compliant. Other global companies such as Hewlett Packard, Monsanto, and Nestlé deployed different versions of their ERP systems in different regional contexts (Davenport, 1998, p. 128). Thirdly, the company lacked sufficient internal IT human resources to design and manage the project. It had to depend almost exclusively on outside consultants. This was extremely expensive. A fourth reason for the increase in cost was the significant effort that had to be put into winning business unit buy-in to the project.

The organizational upheaval, huge cost, and significantly increased complexity led many managers within the company to question the advisability of implementing common information systems enterprise-wide. Given that there was little perceived need for the business units to share day-to-day operational information, business, as well as IT managers, raised questions about whether the effort and expense were worth it. Some felt that a very complex and expensive system had been put in place where something much simpler could have been just as effective. These complex systems now have to be managed, maintained, and upgraded. Instead of achieving economies of scale the organization may have ended up with systems that have created significant diseconomies of scale.

The experience related above should cause both managers and IS researchers to pause. Often managers as well as researchers become quite enamoured with information technology and the latest fad surrounding the technology. Much of current IS research is driven by new technological opportunities (Banville and Landry, 1989). Checkland and Howell (1998) suggest that researchers may be seduced by the glamour of technology. They argue that "many people fascinated by IT are simply not interested in lifting their eyes to such wider questions as what information systems people working in organizations need, how they should be developed, how they should be managed, and what the wider implications of information technology are, both within and between organizations" (p. 55).

Those of us who do research about global information systems should challenge the conventional ideas and assumptions about why and how companies should go about deploying such systems. Given the seductive nature of the technological trends affecting the information systems area, researchers must be willing to delve deeply below the surface of prevailing ideas to raise questions about taken-for-granted assumptions. IS researchers should not be mere prognosticators of the latest information technology fad. They need to critically assess some of the widely held notions embraced by IT consultants and other purveyors of technology solutions. Their research must provide insights as to when and how organizations should deploy integrated global information systems.

Gallupe and Tan's (1999 p. 16) call for more "research into aspects of global enterprise management" is echoed here. Such research should address the following questions among others:

- What are the organizational as well as the technological implications of implementing integrated global information systems?
- Integrated enterprise systems tend to incorporate their own business logic that may require substantial changes in organizational structure and processes. How will these changes affect the competitive positioning and viability of the firm?
- To what extent do global enterprise systems support the underlying strategic business architecture of the global firm?
- To what extent are senior managers willing to commit to and pursue structural as well as cultural organizational change?
- How much integration of global IT systems should companies pursue? Does common information mean implementing the same version of enterprise software in every location?
- How do local cultural, business, and political norms impact global IT systems implementation?

Research focusing on a single variable is inadequate for addressing the issues raised above. By exploring the relationship between multiple variables (Gallupe and Tan, 1999) researchers will be able to provide more significant insights into the opportunities and challenges of implementing global information systems.

REFERENCES

Banville, C. and Landry, M. (1989). Can the field of MIS be disciplined? *Communications of the ACM,* (32), 48-60.

Checkland, P. and Howell, S. (1998). *Information, Systems, and Information Systems: making sense of the field,* Chichester: John Wiley and Sons, 55.

Davenport, T. (1997). *Information Ecology: mastering the information and knowledge environment,* New York: Oxford University Press, 53.

Davenport, T. (1998). Putting the enterprise into the enterprise system, *Harvard Business Review,* July-August, 121-131.

Gallupe, R. B. and Tan, F. B. (1999). A research manifesto for global information management, *Journal of Global Information Management,* 7(3), 5-18.

Jarvenpaa, S. I. and Ives, B. (1993). Organizing for global competition: the fit of information technology. *Decision Sciences*, 24 (3), 547-580.

Kalakota, R. and Robinson, M. (1999). *e-Business: roadmap for success*, Reading, MA: Addison-Wesley, 183.

Chapter 3

Some Internet and E-commerce Legal Perspectives Impacting the End User

Peter P. Mykytyn, Jr.
Southern Illinois University, USA

INTRODUCTION

Not too many years ago, hardly anyone had heard the terms "Web browser," "Web," or "electronic commerce." Today, the World Wide Web, often referred to as simply the Web and as the Internet, offers almost limitless opportunities for end users to do research, obtain comparative information on different products or services, and conduct business online. Many users today, for example, have experienced the opportunity to visit competing web travel sites, e.g., Travelocity.com and Expedia.com, to price airline fares, obtain car rental information, and make a hotel reservation. More often than not, it seems, end users are also intrigued by the fact that prices for the same flight or car are not necessarily the same at the sites searched; in a way, users have become much more savvy in their selection of products and services. In general, end users can become much more efficient and effective as they conduct business online, and both consumers and businesses can participate in unrestricted buying and selling. Consequently, the Web is changing the way businesses do business, and, of course, it is changing the way many end users conduct their business as well.

Electronic commerce (e-commerce) mainly consists of business-to-business (B2B) and business-to-consumer (B2C) types of transactions. According to an e-commerce survey (Survey E-Commerce, 2000) B2B transactions accounted for 80% of all e-commerce and added up to $150 billion in 1999. Further, B2C

Previously Published in the *Journal of EndUser Computing, vol.14, no.1*, transactions in the US amounted to about $20 billion that same year. Although there continues to be a "shaking out" period involving dot.com organizations, questions and decisions about whether to develop Web-based storefronts along with the traditional brick and mortar outlets, e-commerce will most likely continue to expand. But while e-commerce grows, maintaining control over on-line transactions and business risks creates challenges that may not be apparent to unsophisticated end users. One of these challenges pertains to the various and assorted legal issues that confront end users as well as the e-commerce businesses where end users shop. Whether buying or selling on the Web or even just establishing one's home page, legal issues, in addition to providing protection, can also present pitfalls to the unwary.

This paper discusses briefly two of the legal issues that can confront today's end users as they do business over the Web. They are matters dealing with contract law and jurisdictional questions.

CONTRACT LAW ISSUES

Many of the issues dealing with software-related contracts that end users will confront exist because existing contract law is based on the Uniform Commercial Code (UCC), which was written for a goods economy rather than for an economy that is energized by information. In a sense, technology and its use, such as the Web and e-commerce activities, have outpaced the existing contract law perspectives found in the UCC. Zain (2000) suggests that consumer, i.e., end user, protection in e-commerce is needed because of four reasons. First, on-line consumer transactions need facilitation. The lack of regulation in e-commerce has affected on-line purchases and will most likely continue to do so. Examples include concerns over on-line security and privacy, and the enforceability of existing forms of on-line contracts. Second, there is increased ambiguity and risk associated with online sales. In a traditional setting, the end user can visit a retailer and is allowed to browse, see, feel and even smell some merchandise. However, online purchasers lack the certainty and assurance as to what is being purchased and from whom. This type of arrangement was not envisioned with respect to traditional contract law. Zain (2000) also indicates that existing online contracts are inherently unfair to consumers. The reasons include limited or denied warranties, limited remedies, allocating risks to the purchaser who is the least able to absorb such risks, and defining sellers' rights in a very broad and non-reciprocal manner. And, fourth, Zain (2000) suggests that consumers' interest in the enactment of relevant legislation must be safeguarded. Many countries have already proposed and enacted legislation affecting online purchases, and it is important that consumers' interests be represented in the enactment of legislation.

More recently, the Uniform Computer Information Transactions Act (UCITA) was drafted to address the unique aspects of contracts for computer information. UCITA, which was developed by the National Conference of Commissioners on Uniform State Laws (NCCUSL), is a contract law statute that would apply to computer software, multimedia products, computer data and databases, online information, and other such products. It was designed to create a uniform commercial contract law for these products. From the end user's perspective, one of the more familiar perspectives related to UCITA is that it covers contracts that are generally known as "shrink-wrap licenses." It is a law that would have to be ratified by each of the 50 states on an individual basis for it to be effective in a particular state; at the moment only Virginia and Maryland have passed the law, although Virginia imposed certain restrictions when it acted on the law. In 2001, it was introduced in Arizona, Illinois, Texas, Maine, New Hampshire, Oregon, and New Jersey, plus the District of Columbia. The Texas legislature, which adjourned in May 2001, did not pass the law.

Adding to the confusion and possible difficulties end users will be confronted with is the uncertainty about UCITA. It is vigorously opposed by many state attorneys general, the Association for Computing Machinery, the Society for Information Management, software developers, and large software customers. According to Thibodeau (2001), UCITA sets a series of default rules for software licensing transactions, which UCITA's opponents claim are too favorable to software vendors. Opponents charge that UCITA would give software vendors the ability to limit their liability, prohibit reverse engineering, and shut down software remotely in some instances. Conversely, UCITA's proponents argue that the statute is misunderstood and erroneously maligned, with corporate users free to negotiate their own contract terms.

End users, whether those within the corporate sector and/or those who operate as individuals from, say, their home computers should be vigilant about UCITA as it may apply in a user's state. Of course, since the law traditionally lags behind technology and its uses, other contractual legal matters are sure to arise that might impact or confuse end users. For instance, one misperception is that all copies of a contract must be signed and delivered to relevant parties. The Uniform Electronic Transaction Act (UETA), which was passed by the US government in 2000, eliminates legal barriers regarding signed contracts. Instead, electronic signatures, which are defined as being electronic sounds, symbols, email, voice mail, as well as Internet transmissions, are valid substitutes for handwritten signatures. A simple click on "I agree" is enough to show the user's intent. Another misperception is that contracts cannot be made on the Internet because it is too difficult to correct errors. Under UCITA, an electronic contract must provide reasonable means of correcting errors.

JURISDICTIONAL ISSUES

The Internet makes it possible to conduct business throughout the world entirely from a desktop. With this global revolution looming on the horizon, the development of the law concerning the permissible score of personal jurisdiction based on Internet use is in its infant stages (Zippo, 1997).

The following questions are perhaps ones that people do not often ask, but they are certainly relevant in today's cyberspace world of e-commerce. Where can someone be sued? What does the question of jurisdiction mean for the average Internet user, if there is such a person? Since state boundaries are irrelevant on the Internet, how do the courts view jurisdiction where the Internet is concerned? Is there a general consensus among the courts? How can the owner of an Internet home page take steps to see that he/she will not have to defend a lawsuit in a court in a distant state? Jurisdictional issues are becoming increasingly important, as more and more businesses are turning to the Internet to conduct their business, and in turn, more and more end users are doing the same (Lyn, 2000).

Black's Law Dictionary (1990) defines jurisdiction as the "power of the court to decide a matter in controversy and presupposes the existence of a duly constituted court with control over the subject matter and the parties" (p. 853). As far back as 1877, the US Supreme Court dealt with the matter of personal jurisdiction, stating that what was essential was a defendant's physical presence in a forum, i.e., a court or location where cases are usually tried. Since that time, there have been a number of newer interpretations by courts brought about by new technologies, e.g., the telephone, business practices, e.g., telemarketing, and social changes in general. The Internet, the Web, and e-commerce today, as newer technologies and applications thereof, only serve to add confusion to businesses and end users. Perhaps the following scenario, adapted from Lyn (2000) can serve to illustrate the uncertainty.

A small business named "Hollywood Coffee House" located in Somewhere, Maine includes as part of its logo the characters "Emmy Latte" and "Oscar Cappuccino" as a pair of dancing coffee mugs. In an effort to increase business, the business' owner decides to establish a Web presence, i.e., a simple home page, with the Emmy Latte and Oscar Cappuccino highlighted on the Web page. The Web-based approach to increase business is successful, and many new customers from in and around Somewhere, Maine, having seen the home page, now frequent the coffee shop. One day, the business owner is served with complaints issued by the Academy of Television Arts and Sciences and the Academy of Motion Picture Arts and Sciences. Each organization claims that the "Emmy Latte" and "Oscar

Cappuccino" infringes on the respective registered federal trademarks owned by these groups. The complaint alleges trademark dilution and unfair competition. The complaint also requires that the business owner appear in a California federal district court to answer to the complaints. The business owner is most likely confused because he believes his business has nothing to do with California and he himself has never been in California. Reading the complaints further, the owner is told that he is subject to California jurisdiction because the home page is accessible by residents of California and, as such, there are sufficient contacts between the business and California requiring that the owner appear there. As a result of easy-to-use software to develop what may appear to be very unsophisticated Web home pages, business persons like the owner of the coffee shop as well as end users who create their own pages may encounter legal difficulties similar to the scenario described.

The fact that Internet applications are relatively new and that different courts do not all follow the same thinking when dealing with Internet-based personal jurisdiction matters only adds to confusion for businesses and end users. In other words, different interpretations by different courts may result, regarding the coffee house example outlined above. Depending on a court's interpretation of the Internet, one court could find that the business owner's Web page is passive and that the California court would decline personal jurisdiction. Still another court might find that the coffee house's home page is available to users in all states as far as personal jurisdiction is concerned (Lyn, 2000). In a related matter, although the California court may decline personal jurisdiction, the Academy of Television Arts and Sciences and the Academy of Motion Picture Arts and Sciences could still pursue trademark actions against the business owner in Maine. More muddy water to deal with, so to speak!

Although each situation regarding the Internet and personal jurisdiction is obviously different, there are a few guidelines that business persons and end users can follow in order to limit exposure to potential legal difficulties dealing with personal jurisdiction. Avoiding controversy in the first place is one action. For instance, owners of Web pages should conduct trademark searches to ensure that their Web page names do not infringe on someone else's registered trademark. Expressly limiting the reach of a person's or business' Web site by including a prominent statement that the Web site's reach is limited to a particular state or by including a list of states that the Web site owner wishes to avoid may also limit jurisdictional actions. Another suggestion is to limit the use of "autoresponders." An autoresponder is a feature that can be added to a home page that systematically and automatically sends information over the Internet to anyone requesting it through that home page. Similar to the previous suggestion, inquiries from outside a person's state of residence should be addressed carefully, indicating that business activity is not done in the requestor's locality (Lyn, 2000). Schmitt and Nikolai

(2001) also indicate that businesses should review any messages from vendors or customers received through its Web site. Careful scrutiny of customers' physical location is in order. Schmitt and Nikolai (2001) suggest too that businesses watch non-electronic contacts with entities from various jurisdictions. Many cases involving personal jurisdiction and the Internet turn upon whether a defendant had contacts with the plaintiff's forum outside of cyberspace, i.e., through correspondence, telephone calls and physical trips.

CONCLUSION

This paper has dealt briefly with just two areas affecting businesses and end users in today's Web-based e-commerce arena: contractual matters and jurisdictional issues. Other important issues include matters dealing with advertising and intellectual property. With regard to advertising, the Federal Trade Commission (FTC) issued a working paper (Dot Com Disclosures, 2000) confirming that long-established advertising laws, regulations, and guidelines used in traditional media apply equally to online advertising. Even though the working paper cannot be regarded as law, it provides guidelines with respect to how the FTC views Web-based advertising. This means that deceptive or unfair advertising would be illegal and that all advertisement claims must have substantiation. With regard to intellectual property matters, the matter of trademarks was touched on briefly in the paper. However, there is much more to be investigated, i.e., trademark infringement, trademark dilution, invalid use and registration of domain names, improper linking, etc. End users obviously need to be vigilant as they not only surf the Net but conduct business on it.

REFERENCES

Black's Law Dictionary (199). West Group, St. Paul, MN.

"Dot Com Disclosures, *http://www.ftc.gov/bcp/conline/pubs/buspubs/dotcom/index.html*, last accessed June 12, 2001.

Lyn, K.R. (2000). "Personal Jurisdiction and the Internet: Is a Home Page Enough to Satisfy minimum Contacts?," *Campbell Law Review* 22, 341-367.

Schmitt, J. and Nikolai, P. (2001). "Application of Personal Jurisdiction Principles to Electronic Commerce: A User's Guide," *William Mitchell Law Review* 27, 1571-1586.

Survey E-Commerce, Shopping Around the Web, February 26, 2000, *http://www.economist.com/editorial/justforyou/2000226/su7636.html*, last accessed June 12, 2001.

Thibodeau, P. (2001). "UCITA Opponents Slow Software Licensing Law's Progress," *Computerworld*, *http://www. computerworld.com/cwi/story/0,1199,NAV47_STO60652,00.html*, last accessed June 12, 2001.

Zain, S. (2000). "Regulation of E-Commerce by Contract: Is It Fair To Consumers?," *The University of West Los Angeles Law Review* 31, 163-186.

Zippo Mfg. Co. v. Zippo Dot Com, Inc., 952 F.Supp. 1119(W.D. Pa. 1997)

Chapter 4

A New Approach to Evaluating Business Ethics: An Artificial Neural Networks Application

Mo Adam Mahmood, Gary L. Sullivan and Ray-Lin Tung
University of Texas, El Paso, USA

Stimulated by recent high-profile incidents, concerns about business ethics have increased over the last decade. In response, research has focused on developing theoretical and empirical frameworks to understand ethical decision making. So far, empirical studies have used traditional quantitative tools, such as regression or multiple discriminant analysis (MDA), in ethics research. More advanced tools are needed. In this exploratory research, a new approach to classifying, categorizing and analyzing ethical decision situations is presented. A comparative performance analysis of artificial neural networks, MDA and chance showed that artificial neural networks predict better in both training and testing phases. While some limitations of this approach were noted, in the field of business ethics, such networks are promising as an alternative to traditional analytic tools like MDA.

Stimulated by the proliferation of incidents such as tax evasions, defense contractor scandals, insider trading, golden parachutes, executive salaries and bonuses and the savings and loan fiasco, concerns about business ethics have increased significantly over the last decade. Consequently, practitioners and academics are showing increased interest in ethical issues in business. Businesses are updating codes of ethics. Academics are authoring an increasing number of research articles and books.

Research studies have focused on developing theoretical and empirical foundations for understanding the ethics of decision making. Empirical

Previously Published in the *Journal of End User Computing, vol.11, no.3*, studies have used traditional quantitative analytic tools such as multiple regression and multiple discriminant analysis to investigate ethical issues. The present research considers a new procedure, artificial neural networks (ANNs), to analyze ethical decision data. It investigates whether ANNs can outperform discriminant analysis in understanding ethical dilemmas. This comparative test uses ethical judgment data obtained from college students. Using ANNs and discriminant analysis, relationships between these factors and attitudinal variables are assessed.

Several studies of ethical decision making are summarized next. A short presentation of ANNs and discriminant analysis used in analyzing students' ethical perceptions follows. Then, the results of the empirical test are presented along with a discussion of implications. Concluding remarks, including suggestions for future research, complete the paper.

LITERATURE REVIEW

As stated earlier, both public and scholarly interest in business ethics have increased significantly over the past decade (Vogel, 1991). In the next few paragraphs, some recent empirical work in business ethics is reviewed. Empirical work is emphasized in this study because of its centrality to the present research.

This review of empirical studies focuses on business students' and practitioners' judgments regarding ethical issues. For example, DePaulo (1987) examined students' perceptions of the incorrectness of sellers' deceptive bargaining tactics. Interestingly, students were more critical of sellers than were buyers. Claypool, Fetyko and Pearson (1990) compared the responses of CPAs and theologians to ethical dilemmas. When faced with potential ethical dilemmas, both groups indicated that the concepts of "confidentiality" and "independence" were more consequential than "seriousness of breach" and "recipient of responsibility."

Stanga and Turpen (1991) investigated judgments of male and female accounting students on ethical situations relevant to accounting practice. They found no significant gender differences in ethical judgments. Using a nationwide sample of small business employees, Serwinek (1992) investigated the effects of demographic variables such as age, gender, marital status, education, dependent children status, region of the country and years in business on ethical perception. Age was found to be the most significant factor in predicting ethical perception. Premeaux and Mondy (1993) used marketing managers to investigate the link between ethics and management behavior. The authors established that, even with recent heightened concerns for ethical issues in business, this link has not changed much since the mid-

1980s. When making business decisions, practitioners still depend on utilitarian ethical philosophy.

Galbraith and Stephenson (1993) studied business policy students to investigate whether males and females use different decision rules when making ethical value judgments. The authors found that there are situations where genders use different decision rules and situations where they use the same rules.

In the past, empirical ethics research studies used traditional statistical tools such as discriminant analysis, factor analysis, cluster analysis, regression analysis and Linear Structural Relation (LISREL) analysis. For example, Cohen, Pant and Sharp (1993) used factor analysis and regression analysis to validate a multidimensional ethics scale proposed by Reidenbach and Robin (1990). Souter, McNeil and Molster (1994) used discriminant analysis and cluster analysis to examine ethical conflict experienced by employees in Western Australia. Akaah and Lund (1994) used a LISREL model to investigate the influence of personal and organizational values on marketing professionals' ethical behavior.

The present research begins the analysis and discussion of ANNs' applicability in the field of business ethics. Even though prior research in the area has addressed ethical issues concerning AI (Khalil, 1993; Dejoice, Fowler and Paradice, 1991; Mason, 1986) no study has yet used ANNs to study ethical decisions. Additional considerations favoring ANNs versus traditional tools like discriminant analysis include: 1) ANNs improve by using one data element at a time (increasing learning potential) while discriminant analysis considers all training data simultaneously. 2) ANNs do not require *a priori* model specification (increasing the potential for serendipity in exploratory studies) while discriminant analysis needs an *a priori* model. 3) Lippman (1987) has suggested that ANNs may be more robust than alternative models.

Because ANN is untested in ethics research, it is difficult to predict performance compared to more established methodologies. Because ANN appears to have great, if untested, potential for ethical inquiry, ANN was pitted against multiple discriminant analysis (MDA) to compare predictive potential. Relationships between students' ethical judgments and their attitudes were investigated. The context of this research is timely on content grounds because "researchers within the field of business ethics are quite interested in business students' and practitioners' ethical judgments regarding ethical issues within business" (Barnett, Bass and Brown, 1994; p. 470). The methodological dimension of the present research also contributes by providing procedures to design, train, and test ANNs for ethical decision making. Future researchers interested in using ANNs in ethical studies should find these helpful.

RESEARCH METHODS

Artificial Neural Networks

Artificial neural networks (ANNs) are information processing tools. An ANN consists of processing elements called neurons (also known as neurodes). Neurons communicate with each other through weighted paths called connections. Each connection can be characterized as excitatory or inhibitory (Zahedi, 1993). An excitatory connection from a neuron increases the strength of a receiving neuron, while an inhibitory connection reduces its strength. The network learns connection weights through a training process where cases from a training set are repeatedly fed to the network.

Neurons in an ANN are organized in layers. The first layer is called the input layer, and the last layer is known as the output layer. The middle layer is designated as the hidden layer. There can be more than one hidden layer, even though the most common type of neural network consists of one hidden layer. Through the repetitious feeding of training examples, the input layer neurons receive facts about a decision problem or an opportunity. These neurons, as a result, become energized and send outputs to neurons in the hidden layer(s). The hidden layer(s)' neurons facilitate generalizations on the part of the network by transforming externally derived input facts into higher level features. These results are then sent to neurons in the output layer. The output layer neurons, in turn, communicate system results to the user.

Each neuron has an activation level. Activation level refers to the strength of the neuron. It is derived by a linear or nonlinear function associated with the neuron that combines incoming connection weights to the neuron into a single output. The design of an artificial neural network for a given problem consists of deciding on (Zahedi, 1993): a network topology (i.e., number of neuron layers as well as interlayer and intralayer connections), an activation function for the nodes to convert inputs to outputs and a training process to allow the network to learn from the training examples fed. The network designed for the present research is a fully connected feedforward backpropagation network with three layers: an input layer, an output layer, and a hidden layer. deVilliers and Bernard (1992) investigated the classification and training performance of fully connected feedforward backpropagation artificial neural networks with one and two hidden layers. Their conclusion was that artificial neural networks with two hidden layers do not perform better than artificial neural networks with one hidden layer.

Cybenko (1989) and Funahashi (1989) also found that one hidden layer is enough to approximate any continuous function.

In the network designed for the present research, each layer is fully connected to the succeeding layer and there are no intralayer connections. The fully connected configuration means that each neuron on each layer is connected to every neuron on the succeeding layer. The present network also uses feedforward connection, in which the neurons on each layer send their output to neurons on the succeeding layer. Thus, data flows are all in one direction and there are no feedback loops from a neuron or layer to a previous one.

The artificial neural network designed for the present research uses the backpropagation learning algorithm (Rumelhart, Hinton and Williams 1986). Backpropagation networks with feedforward connections have been identified as highly efficient categorizers (Waibel et al., 1989; Barnard and Casasent, 1989; Burr, 1988). Though it does not ensure an optimal solution, solutions generated through the algorithm are found to be close to optimal (Rumelhart, Hinton and Williams, 1986).

Backpropagation is a gradient descent-based algorithm which uses a training sample to arrive at an optimal output. During the training phase, the network is provided with the input and desired output which is compared against the actual output generated by the network. If any difference exists between the actual and desired output, the network adjusts the relevant connection weights to reduce this difference and, in the process, uncovers the pattern underlying the relationship between the input and the desired output (Kirrane, 1990). The artificial neural network designed for the present research uses the generalized delta rule to adjust connection weights because for most backpropagation networks it is the "common choice" (Using NWorks ,1991).

Another issue in building an artificial neural network is the type of activation function used in generating the output in a neuron. The network designed for the present research uses the sigmoid activation function. The sigmoid transfer function is used to adjust connection weights associated with each input neuron. It is common to use the sigmoid transfer function with the backpropagation learning algorithm (Using NWorks 1991).

The software package, NeuralWorks Professional II/PLUS developed by Neuralware, Inc., running on 486 IBM compatible personal computer was used for designing, training and testing the artificial neural networks used in this study. This particular package is used because it provides a complete artificial neural network development and deployment environment (Using NWorks, 1991; p. 7). This system is especially appealing because it allows a fast and easy way to create a backpropagation network and modify its learning parameters during the training phase which may influence the network's learning time and performance (Reference Guide to NeuralWorks Professional II/Plus 1991). Once trained, the system allows the user to view connection weights and outputs related to the network.

Instrument

Because this research is exploratory and focuses on the application of a new analytic tool, with theory development a secondary consideration, no reliability and validity tests were conducted on the instrument which measured students' attitudes and reactions to ethical situations. There were three parts to the questionnaire. Part I of the instrument consisted of six demographic questions (see Appendix A). Three open-ended questions assessed age, major and high school location. Three closed-ended items sought information on gender, student status and work experience.

Part II of the instrument presented respondents with four situations (e.g., SQ1, SQ2, SQ3 and SQ4) faced by consumers which may have ethical content. These situations were developed following a review of relevant literature. The intent was not to include all ethical situations faced by consumers but to get a representative sample of such situations. Clearly, it would be impossible to cover the complete range of ethical situations faced by consumers within a single research instrument.

Part III asked participants six attitudinal questions (e.g., SQ5 (has 15 sub-questions), SQ6, SQ7, SQ8, SQ9 and SQ10). These questions sought insight into students' attitudes toward different groups of people, businesses and work environments.

The Data Set

One hundred-sixty responses to a survey investigating the ethical perceptions of students attending a western U.S. college were gathered. Participants were asked to evaluate the ethics of a variety of groups and to forecast their own behavior in various ethical situations.

Four ethical situations (depicted by SQ1, SQ2, SQ3 and SQ4) were analyzed. Responses to SQ5, SQ6, SQ7, SQ8, SQ9 and SQ10 were used as inputs to the MDAs and ANNs designed for the present research. Responses to SQ1, SQ2, SQ3 and SQ4 served as outputs for ANNs and MDAs. The objective was to investigate the possible relationship between students' perceptions of an ethics-related situation and their attitudes toward different people, businesses and work environments.

Before it can be used in MDA and to design, train and test an ANN, a data set must be carefully evaluated. Generally, good data contains representative samples of sufficient examples and known inputs and outputs for each example before these can be used in the design of an ANN. In other words, for the network to comprehend an ethical decision situation, the data set must contain sufficient examples in each of its subsamples (Lawrence and Andriola, 1992). Following this general rule, categories with ten or fewer observations were eliminated. This left 6, 5, 3 and 5 subcategories under SQ1, SQ2, SQ3 and SQ4, respectively.

ANNs create their unique classifying rules by training on part of the available data. These rules are then tested on the rest of the data. A similar practice is followed for MDAs. A part of the available data set is used to create initial MDA models. Following model estimation on this data, models are tested using the remaining data. To facilitate this process, the data set under each of the ethical situations (e.g., SQ1, SQ2, SQ3 and SQ4) was randomly divided into two equal size groups, a training group and a testing group.

Training the Artificial Neural Networks

In this research, a set of backpropagation ANNs with feedforward connections has been developed. The inputs to the ANNs will be business students' attitudes toward a variety of people, businesses, and work environments and the outputs from ANNs will be their judgments regarding ethical issues within business. More specifically, using the aforementioned software, an ANN was designed for each set consisting of responses to SQ5, SQ6, SQ7, SQ8, SQ9 and SQ10 as inputs and responses to SQ1, SQ2, SQ3 or SQ4 as outputs. Supervised learning (in which the network was told the correct answer) was conducted to train the networks. Because the total subcategories for SQ5 to SQ10 were 85 (see Appendix B), the number of nodes in the input layer, for all networks, was specified as 85. And because the subcategories under SQ1 through SQ4 were 6, 5, 3 and 5, the numbers of nodes in the output layer were 6, 5, 3 and 5 for SQ1 through SQ4, respectively.

Unfortunately, there is no definitive method (since artificial neural networks are still under development) to determine the number of nodes in the hidden layer. This is done by trial and error. The present research used the following procedure. For each network, the number of output nodes was selected as the initial number of hidden nodes. One node was added in each of the next eleven steps. Because each of SQ1 to SQ4 has different output nodes, the ranges attempted for each data set were various.

The numbers tried were:

Set	Nodes in Hidden Layer
SQ1 :	6, 7, ..., 17
SQ2 :	5, 6, ..., 16
SQ3 :	3, 4, ..., 14
SQ4 :	5, 6, ..., 16

The objective was to select, for each set, the network with the highest accuracy rate. It is possible that multiple networks (having different numbers of nodes) will tie at the highest accuracy rate. If so, the best network would be the one with fewer nodes, as it takes less time to train a network with fewer hidden nodes and is less sensitive to available degrees of freedom.

Further attempts were made to improve the results obtained through the networks. For example, the networks having the best performances in the previous step were selected and then the number of hidden nodes was doubled at each step (subject to the maximum imposed by the number of nodes in the input layer) until classification error was no longer reduced (Marquez et al. 1991). This resulted in no further improvement in results. Another approach suggested by Salchenberger, Cinar and Lash (1992) was also tried. They suggested that the number of nodes in the hidden layer should be 75% of the number of nodes in the input layer. This formula resulted in 64 as the theoretical number of nodes in the hidden layer for all four ethical decision artificial neural networks (recall that the number of nodes in the input layer was 85). Using 64 hidden nodes failed to improve results.

Next, a sequence of advanced refinements on the previous results was attempted by varying momentum, learning coefficient, epoch, and threshold. The effectiveness and convergence of the backpropagation learning algorithm depend on the value of these parameters. The momentum and learning coefficients were used to accelerate the convergence of the backpropagation learning algorithm. The momentum term represents the proportion of the weight adjustment attributed to the recent classification error. Typically, the momentum term is chosen between 0.1 and 0.8 (Zurada, 1992). The learning coefficient determines the effect of past weight changes on the current direction towards successful convergence of the backpropagation learning algorithm. There is no single learning coefficient value suitable for different training cases (Zurada, 1992). This value depends on the problem being solved.

Momentum and learning coefficient default values are given by NeuralWorks Professional II/PLUS. These were used as the starting point.

Accordingly, the learning coefficient and momentum were set at .5 and .4, respectively. Both the learning coefficient and momentum were adjusted upward and downward during the training phase in an effort to improve performance. These adjustments introduced no improvements and, in fact, impaired the networks' performance.

Threshold refers to a convergence threshold which, when reached, is used to stop learning. Epoch, on the other hand, is the number of sets of training data (learning cycles) presented to the network between weight updates. Training begins with arbitrary values for the weights (they might be random numbers) and proceeds iteratively. Each iteration is called an epoch. In each epoch, the network adjusts the weights in the direction that reduces error (the difference between the current outputs and the target outputs). As the iterative process of incremental adjustments continues, weights gradually converge on the optimal set of values. Usually, many epochs are required before training is complete (Smith 1993).

Default values of threshold and epoch are given by NeuralWorks Professional II/PLUS. These were used as the starting point. Accordingly, threshold and epoch were set at .05 and 16, respectively. Both threshold and epoch were adjusted upward and downward during training to improve the performance of the network. These adjustments produced a slight improvement in classification rates, ranging from 1.3 percent to 2.9 percent. The final network configurations are presented in Table 1.

Table 1: Final Configurations of ANN Models

	Input PEs	Hidden PEs	Output PEs	Epoch	Threshold	Momentum	Learning coefficient
SQ1	85	6	6	16	0.02	0.4	0.5
SQ2	85	5	5	32	0.05	0.4	0.5
SQ3	85	9	3	16	0.02	0.4	0.5
SQ4	85	9	5	16	0.05	0.4	0.5

EMPIRICAL RESULTS

Table 2 presents the comparative performance of ANNs on the training data and the test data. These results indicate that ANNs had 99.32 percent mean predictive accuracy for the training data set versus 48.48 percent for the test data.

Table 4 presents the results of MDAs. These results show that MDAs had 62.44 percent mean predictive accuracy compared to 32.35 percent for the test data. Although the accuracy rate for ANNs fell sharply under test conditions, the ANNs' accuracy was significantly higher than chance levels (see Table 3).

Table 2: Actual Performance of ANN

	SQ1	SQ2	SQ3	SQ4	Mean
Training Phase	98.67	98.61	100	100	99.32%
Testing Phase	36.0	41.7	70.6	45.6	48.48%

Table 3: Chance Performance of ANN

	SQ1	SQ2	SQ3	SQ4	Mean
Chance Probability	16.67	20.00	33.33	20.00	22.50

Table 4: Actual Performance of MDA

	SQ1	SQ2	SQ3	SQ4	Mean
Training Phase	57.33	58.33	72.06	62.03	62.44%
Testing Phase	29.33	26.39	38.24	35.44	32.35%

DISCUSSION

The present exploratory research suggests that results obtained through artificial neural networks (ANNs) are superior to both the chance criterion and the multiple discriminant analysis (MDA) approach in both the training phase and the testing phase of model development. More specifically, in predicting students' judgments regarding ethical issues within business, ANNs outperformed MDAs for all four data sets. In the training phase, ANNs exceeded MDA's mean predictive accuracy by about 37 percentage points. In the testing phase, ANNs topped MDA's mean predictive accuracy by over 16 percentage points. More importantly, ANNs outscored the chance probability by about 68 percentage points in the training and fully 17 points in the testing phase of model development.

As stated earlier, the mean accuracy rate for test data for ANNs is 49 percent of the level attained using the training data (the MDA's decline was comparable). There are many possible causes for this decline in predictive power. First, 80 cases of training dataset and 80 cases of testing dataset (with 85 network input nodes) may not adequately represent the sampling population. Lawrence and Andriola (1992) stated that the data must include examples of sufficient variety for a network to generalize. Second, in contrast to other ANN applications (e.g., banking, bankruptcy, and stock prediction), ethical perception is subjective. The subject matter may have caused contradictions and ambiguities in the data which lessened the networks' predictive ability. Third, the degrees of freedom were highly constrained by the quantity of available cases.

While the findings of the present research are preliminary, it appears that ANNs may be more helpful in developing accurate predictive models in business ethics than traditional analytic tools like MDA. The investigators believe that the ANN methodology holds promise as a tool for empirical research in the field of business ethics. More specifically, ANNs have important research advantages including the ability to adjust rapidly to reflect changes in the real world. And, unlike MDA, networks do not require that data be normally distributed or have equal variance.

Unfortunately, current applications of ANN technology have several limitations. An inherent problem of ANNs is that there are no formal guidelines for selecting a network configuration for a given prediction task. Without guidelines, the selection of structures and parameters will continue as a trial and error process. It is hoped that more systematic guidelines will be developed soon.

Explanatory capability is another problem associated with ANNs. Once trained, ANNs are essentially considered "black boxes" even though there are methods in the literature which try to extract rules for a network. The internal structure of the black box makes it difficult to trace the steps by which the output is reached. The information is contained in its nodes, links and weights which cannot be translated into an algorithm that would be intelligible or useful outside the ANN. Thus, there is no way to verify if the network's structure conforms with theorized concepts and causal relationships. The only way to evaluate the network for consistency and reliability is to monitor its output. This may not be a serious disadvantage when the concern is mainly prediction, as in the present ethical perception study. But, if the characteristics of each group and the significance of each input are a concern, the present technology may not be adequate.

Finally, networks require advanced hardware and software. And training an ANN demands more computation time than other classification methods.

CONCLUSIONS AND RECOMMENDATIONS FOR FUTURE RESEARCH

In this exploratory research, we have presented a new approach to classifying, categorizing and analyzing ethical decision situations. A comparative performance analysis of ANNs versus MDA and chance indicated that artificial neural networks are better predictors in both training and testing phases. While some limitations of this approach were noted, in the field of business ethics, these networks possess considerable potential as an alternative to traditional analytic tools like MDA. More empirical research studies must be conducted before the full potential of ANNs can be established in the ethics field.

Future research should utilize carefully designed and validated instruments for data collection and larger samples for more stable results. Techniques such as factor analysis and stepwise multiple discriminant analysis should be used to purify input data (to reduce noise) prior to using ANNs to analyze ethical decision making. Later studies should seek input from practitioners as well as students to broaden understanding of issues in business ethics.

ACKNOWLEDGMENT

The authors thank Dr. P.J. O'Connor, Visiting Associate Professor of Marketing at Loyola Marymount University, for allowing the use of survey data referenced in this study.

REFERENCES

Akaah, I.P. and Lund, D. (1994). Influence of Personal and Organizational Values on Marketing Professionals' Ethical Behavior. *Journal of Business Ethics,* 13, 417-430.

Barnard, E. and Casasent, D. (1989). A Comparison between Criterion Functions for Linear Classifiers with an Application to Neural Nets. *IEEE Transactions on Systems, Man, Cybernetics,* 1030-1041.

Barnett, T., Bass, K. and Brown, G. (1994). Ethical Ideology and Ethical Judgment Regarding Ethical Issues in Business. *Journal of Business Ethics,* 13, 469-480.

Burr, D.J. (1988). Experiments on Neural Net Recognition of Spoken and Written Text. *IEEE Transactions of Acoustic, Speech,* Signal Processing, ASSP-36, 1162-1168.

Claypool, G.A., Fetyko, D.F., and Pearson, M.A. (1990). Reactions to Ethical Dilemmas: A Study Pertaining to Certified Public Accountants. *Journal of Business Ethics,* 9, 699-706.

Cohen, J., Pant, L. and Sharp, D. (1993). A Validation and Extension of a Multidimensional Ethics Scale. *Journal of Business Ethics,* 12, 13-26.

Cybenko, G. (1989). Approximations by Superimpositions of Sigmoidal Functions. *Math, Control, Signals, and Systems,* 2, 303-314.

Dejoice, R., Fowler, G., and Paradice, D.(1991). *Ethical Issues in Business Information Systems.* Boston, MA: Boyd and Fraser Publishing Company..

DePaulo, P.J. (1987). Ethical Perception of Deceptive Bargaining Tactics Used by Salespersons and Consumers: A Double Standard. *Proceedings of the Division of Consumer Psychology* (American Psychological Association), 201-203.

deVilliers, J. and Barnard, E. (1992). Backpropagation Neural Network with One and Two Hidden Layers. *IEEE Transactions on Neural Networks,* 4(1), 136-141.

Funahashi, K. (1989). On the Approximate Realization of Continuous Mappings by Neural Networks, *Neural Networks*, 2, 183-192.

Galbraith, S. and Stephenson, H.B. (1993). Decision Rules Used by Male and Female Business Students in Making Ethical Value Judgments: Another Look. *Journal of Business Ethics,* 227-233.

Khalil, O.E.M. (1993). Artificial Decision-Making and Artificial Ethics: A Management Concern *Journal of Business Ethics,* 12, 313-321.

Kirrane, D.E. (1990). Machine Learning. *Training & Development Journal,* 24-29.

Lawrence, J. and Andriola, P. (1992). Three-Step Method Evaluates Neural Networks for Your Application, *EDN,* August 6, 93-100.

Marquez, L., Hill, T., Worthley, R., and Remus, W. (1991). Neural Network Models as an Alternative to Regression, *Proceedings of the Twenty-Fourth Annual Hawaii International Conference on Systems Sciences*, IV, 129-134.

Mason, R.O. (1986). Four Ethical Issues of the Information Age. *MIS Quarterly,* 10, 5-12.

Premeaux, S.R. and Mondy, R.W. (1993). Linking Management Behavior to Ethical Philosophy, *Journal of Business Ethics*, 12, 349-357.

Reference Guide: NeuralWorks Professional II/Plus and NeuralWorks Explorer. (1991). Pittsburgh, PA: NeuralWare, Inc..

Reidenbach, R.E. and Robin, D.P. (1990). Toward the development of Multidimensional Scales for Improving Evaluation of Business Ethics. *Journal of Business Ethics,* 639-653.

Rumelhart, D.E., Hinton, G. and Williams, R. (1986). *Learning Representation by Back-Propagation Errors. Nature,* 9, 533-536.

Salchenberger, L.M., Cinar, E.M., and Lash, N.A. (1992). Neural Networks: A New Tool for Predicting Thrift Failures. *Decision Sciences*, 23, 899-916.

Serwinek, P.J. (1992). Demographic and Related Differences in Ethical Views Among Small Businesses. *Journal of Business Ethics,* 11, 555-566.

Smith, M., (1993). *Neural Networks for Statistical Modeling,* New York, NY: Van Nostrand Reinhold.

Souter, G., McNeil, M.M. and Molster, C. (1994). The Impact of the Work Environment on Ethical Decision Making: Some Australian Evidence. *Journal of Business Ethics,* 13, 327-339.

Stanga, K.G. and Turpen, R.A. (1991). Ethical Judgments on School of Accounting Issues: An Empirical Study. *Journal of Business Ethics,* 10, 739-747.

Using Nworks: An Extended Tutorial for NeuralWorks Professional II/Plus and NeuralWorks Explorer, (1991). Pittsburgh, PA: NeuralWare, Inc.

Vogel, D. (1991). New Perspectives on Old Problems. *California Management Review*, 101-117.

Waibel, A., Hanazawa, T., Hinton, G., Shikano, K. and Lang, K.J. (1989). Phoneme Recognition Using Time Delay Neural Networks. *IEEE Transactions on Acoustics, Speech, Signal Processing,* 37, 328-339.

Zahedi, F. (1993). *Intelligent Systems for Business: Expert Systems with Neural Networks*. Wadsworth Publishing Company, Belmont, CA.

Zurada, J. M.(1992). *Introduction to Artificial Neural System,* West, St. Paul, MN.

APPENDIX: GRADUATE RESEARCH STUDENT ETHICS SURVEY QUESTIONS

Your Age: Sex: M F

Major:

Check one: Grad Undergrad

Where did you graduate from high school (state or country)?

How would you describe the majority of your work experience?
A. Part-time
B. Full-time
C. No work experience

Please answer the following questions honestly. Circle the choice that describes what YOU WOULD DO in the situation, not what you think is right to do:

SQ 1. You go to a university that has an arrangement allowing students to buy IBM software at a 35% discount provided they sign a written certification that it is for personal use. A good friend asks you to buy some software for him at a discount. He will pay you. He says he needs the software but that he cannot afford to pay the full price ($295). He says if he can't get it at a discount, he will find someone who has the software and make an illegal copy. You know that many students buy software for friends and relatives and you assume the school knows this. You would:

a. Buy it for him. It is a common practice and if IBM or the school wanted to stop it, they would have developed better safeguards.
b. Buy it for him. Since you would have bought the software for yourself, neither the school nor IBM loses anything.
c. Buy it only if you believe your friend really needs it and can't afford to pay the full price, though you think it is wrong to say it is for your personal use when it isn't.
d. Buy it though you think it is wrong to do so, since it would be worse if your friend made an illegal copy; at least this way IBM gets something.
e. Make up an excuse as to why you can't buy it for him; for example, tell him that there are additional requirements about the software relating to a current course.
f. Tell him that you won't do it, but try to help him get an illegal copy from someone who has the software.
g. Tell him politely that you won't do it because using the discount in this way is wrong.
h. Tell him politely but firmly that you won't lie for him and that you think it would be wrong to make an illegal copy.

SQ 2. Your car is rear-ended by another car, damaging your rear bumper. The other driver is insured. When you go to a body shop for an estimate, the estimator suggests that he can also fix a rear fender dent that you had before the accident. He says that you can claim that the damage was caused by the recent collision. Otherwise, fixing the fender will cost $375. He assures you that he has done it many times before and that you will have no trouble with the insurance company. You would:

continued on next page

a. Include the fender in your claim if you think your insurance rates are too high and you have not had any previous claims.
b. Include the fender in your claim since the estimator suggested it and it is apparently an expected and common practice.
c. Include the fender in your claim because it is an unexpected opportunity and you could not otherwise afford to fix the fender, though you think it is wrong to do so.
d. Politely decline the opportunity and only put in a claim for the actual damage because no matter what he says you could get caught and be accused of fraud.
e. Politely decline the opportunity and only put in a claim for the actual damage because it would be wrong to file a false claim.
f. Firmly decline the suggestion and tell him that you think it is wrong.
g. Firmly decline the estimators' suggestion and report him to the insurance company.

SQ 3. You work for a large toy manufacturer. Two months before Christmas you discover that your company's best-selling toy has a defect, making it potentially dangerous to children. Your boss says the risk of injury is small and that a recall is out of the question. You disagree. He adds that your job could be in jeopardy if you pay further attention to the situation. What would you do?

a. Ignore the situation and hope for the best.
b. Write a memo outlining your concern to your boss and your boss's superior, suggesting that the toy be recalled.
c. Quit your job.
d. Make a confidential call to the federal Consumer Products Safety Commission (CPSC) and tell them what is going on.
e. Make a confidential call to your local newspaper and tell them what is going on.
f. Ignore the situation for now, but work toward change in the future.

SQ 4. You work in a foreign office of an American manufacturer of heavy equipment. One of your goals is to get a substantial contract from the local government. The competition is fierce and to get the contract , you've been told to pay off several foreign officials. What would you do?

a. Make the payoffs.
b. Refuse to participate.
c. Quit and get another job.
d. Ask to be transferred to the United States.
e. Other.

SQ 5. How do you rate the overall honesty and ethics of the following groups:

5 = Excellent
4 = Very Good
3 = Good
2 = Bad
1 = Very Bad

		CIRCLE ONE				
1.	Elected public officials	1	2	3	4	5
2.	Successful business executives	1	2	3	4	5
3.	Journalists	1	2	3	4	5
4.	Judges	1	2	3	4	5
5.	Lawyers	1	2	3	4	5

continued on next page

6.	Police Officers	1	2	3	4	5
7.	Famous Athletes	1	2	3	4	5
8.	Famous Musicians	1	2	3	4	5
9.	Teachers at Your College	1	2	3	4	5
10.	Your Parents	1	2	3	4	5
11.	Students at Your College	1	2	3	4	5
12.	Your Friends	1	2	3	4	5
13.	People over 30	1	2	3	4	5
14.	People under 30	1	2	3	4	5
15.	Yourself	1	2	3	4	5

SQ 6. Would you invest in a company with substantial holdings in South Africa?

1. Yes
2. No

SQ 7. If you were employed by a company that produced/manufactured goods that were harmful to human beings, would you quit?

1. Yes
2. No

SQ 8. Is it okay to put your personal mail (1 or 2 pieces) in the company mail basket?

1. Yes
2. No

SQ 9. It is okay to have a company pay for your education knowing that you don't plan on staying with them?

1. Yes
2. No

SQ 10. Would you hire a friend of yours, who was qualified for a job you had open, even though there are applicants that are more qualified?

1. Yes
2. No

Chapter 5

Copyright, Piracy, Privacy, And Security Issues: Acceptable Or Unacceptable Actions For End Users?

Jennifer Kreie and Timothy Paul Cronan
University of Arkansas

End user acceptable/unacceptable behavior related to computer information systems has caused significant losses to business and society. Some measures have been suggested to prevent losses or discourage unethical behavior. One approach is to identify factors that might better explain acceptable/unacceptable actions. In this study, we examined models of ethical behavior and selected factors from these models that might influence decision-making. We conducted an empirical study to determine perceptions of acceptable or unacceptable behavior and which factors were significant in influencing a person's judgment of acceptable/unacceptable behavior. Identifying the factors that influence ethical decision-making is one approach suggested by researchers to find ways to discourage unethical behavior and help businesses prevent losses. The results of the study indicate there are significant factors that affected individuals' assessment of what is acceptable behavior. The results also indicate that the factors which were influential were contingent on the characteristics of the ethical dilemma.

Computer-based information systems are ubiquitous in business and are becoming so throughout society. The proper use of computers has been beneficial to businesses, but misuse, and unethical behavior related to

Previously Published in the *Journal of End User Computing, vol.11, no.2*, Copyright © 1999, Idea Group Publishing.

information systems (IS) have caused significant losses to businesses and society. Fortune magazine (February 1997) recently reported on an actual breach of security to illustrate how easily a business' information system could be accessed and misused.

Estimates for the annual cost of computer crime are in the billions. Though organizations have invested in the development and implementation of security measures, computer misuse will likely continue to be a problem throughout the 1990s (Straub and Nance, 1990). From a professional and social perspective, IS professionals are concerned about the problem of illegal and/or unethical use of computers because of the potential harm to society and to the integrity of the IS profession. There has been research in the area of computer misuse and computer ethics in recent years from several perspectives, such as the ethical attitudes of personnel, deterrents to unethical behavior, the various types of unethical behavior, and approaches to teaching ethics in the field of MIS (Zalud, 1984; Saari, 1987; Aiken, 1988; Heide and Hightower, 1988; Cougar, 1989; Oz, 1990; Paradice, 1990; Straub and Nance, 1990; Conner and Rumelt, 1991).

In an effort to prevent or discourage improper and unethical use of computers several measures have been suggested—enhanced security, prompt and fair reporting, tougher sanctions, and codes of conduct for IS professionals (Anderson, et al., 1993; Straub, 1990). It would also be beneficial to identify the ethical issues that pertain to computer use and include ethics in the IS curriculum (Parker, 1980; Cougar, 1984). This would help to make future IS professionals more aware of ethical issues. A better understanding of what factors influence our decision to act in one way or could, perhaps, provide some guidance for discouraging unethical behavior. For example, it would be beneficial to know whether a code of ethics for IS professionals has a significant impact on a person's decisions and behavior.

The current study is part of an ongoing research effort to identify factors which affect the ethical behavior of individuals when these individuals are faced with ethical dilemmas involving different aspects of computing. The purpose of the present study is to examine factors proposed in several models of thical/unethical decision-making, several of which have not been empirically tested, and determine what factors are significant in the judgment of what is acceptable or unacceptable behavior. In the present study, participants were presented with an IS dilemma and a response. Participants were asked to judge the response as "acceptable" or "unacceptable" and to indicate what factors influenced their decision.

ETHICAL BEHAVIORAL MODELS

Several models have been suggested that relate specifically to the issue of ethical judgments and behavior. Most of the models described below have not been empirically validated. Rest (1979) proposed a four-component model to explain ethical behavior. The four components of the model interact to explain the psychology of moral judgment—realization, decision, interconnecting cognition and affect, and execution. Ferrell and Gresham (1985) proposed a model of the decision-making process in which they categorized variables that affect the process into individual factors (knowledge, values, attitudes, and intentions), significant others (differential association and role set configuration), and opportunity (professional codes, corporate policy, and rewards/punishment).

Using Kohlberg's (1969) moral development stages and Simon's (1960) intelligence phase, Trevino (1986) posed a model for explaining ethical behavior in organizations. Trevino model includes individual moderators (ego strength, field dependence, and locus of control) and situational moderators (immediate job context, organizational culture, and characteristics of the work).

Eining and Christensen (1991) examined the issue of software piracy and proposed a psycho-social model to explain software piracy behavior. Their model incorporates computer attitudes, material consequences, norms, socio-legal attitudes, and affective factors as independent variables that impact software piracy. They suggested that these variables contribute to intentions which, in turn, lead to behavior. Eining and Christensen reported empirical results that all these variables, except socio-legal attitude, are significant in explaining the variation in software piracy behavior.

Bommer, Gratto, Gravander, and Tuttle (1987) proposed a model to explain general ethical behavior. Their model consists of several variables which influence the decision process (information acquisition, information processing, cognitive process, perceived rewards, and perceived losses) to determine ethical or unethical behavior. These factors include individual attributes (moral level, personal goals, motivation mechanism, position/status, self concept, life experiences, personality, and demographics), personal environment (peer group and family), professional environment (codes of conduct, licensing requirements, and professional meetings), work environment (corporate goals, stated policy, and corporate culture), government/legal environment (legislation, administrative agencies, and judicial system), and social environment (religious values, humanistic values, cultural values, and societal values).

Finally, there are more general models of behavior that have been applied to ethical decision-making and behavior. The theory of reasoned action (TRA) proposed by Ajzen and Fishbein (1980) states that *intention* precedes behavior and two factors directly affect intention — attitude toward the behavior and subjective norm. The *intention* of a person to perform a behavior (such as in the context of an ethical dilemma) is a good indicator that the person will perform the behavior. The "attitude toward the behavior" directly affects *intention*, and *attitude* reflects the degree to which a person has a favorable or unfavorable evaluation of the behavior. The "subjective norm," which is the perceived social pressure to perform or not to perform a behavior, also affects *intention*. Schwartz and Tessler (1972) presented a modification to TRA model in which they added personal normative beliefs as a direct influence on the *intention* to perform a behavior. Personal normative beliefs reflect the sense of obligation a person feels about performing a behavior (Schwartz and Tessler, 1972; Beck and Ajzen, 1991). Schwartz and Tessler (1972) reported empirical support that the measure of personal normative beliefs was the strongest predictor of a behavior.

METHOD

The models previously described propose several factors that may affect how a person evaluates an ethical situation and judges behavior. The purpose of this research was to determine whether significant factors can be identified which influence ethical decision-making. For a given ethical dilemma, an individual is hypothesized to use several factors as the basis for deciding what is ethical or unethical. The suggested factors are *societal environment, belief system, personal values, personal environment, professional environment, legal environment, business environment, moral obligation,* and *awareness of consequences*. Table 1 summarizes these factors.

Seven of the factors can be related to the factors in the Bommer et al. (1987) model, although the names for the factors are not always identical. The *societal environment* represents the social and cultural values that impact the individual. The *belief system* represents the religious values and beliefs within one's spiritual or religious environment. *Personal values* represent the personal experiences, goals, and moral level of a specific individual and *personal environment* represents the influence of family, peers, and significant others (this factor was also suggested by Ferrell and Gresham [1985]). The *professional environment* represents the codes of conduct and expectations within a profession. The *legal environment* represents the law and government. The *business environment* reflects the corporate goals and profit

Table 1: Factors Proposed to Influence Ethical Decision-Making

FACTOR	DESCRIPTION
Environment	
societal	Social/cultural values. *What does society say should be done?*
belief system	Religious values and beliefs from one's spiritual or religious environment. *What does one's church/religion say?*
individual	Significant others; peer group. *What does mom or my close friend say?*
professional	Codes of conduct and professional expecta tions. *What does my profession say?*
legal	Law and legal issues. *What does the law say?*
business	Corporate goals and profit motive. *What does my company and the "bottom line" say?*
Personal values	One's internalized values and experiences. *What do I say?*
Characteristics of the individual	Gender, age, education, work experience, etc.
Moral obligation	A feeling of responsibility or obligation.
Awareness of consequences	Association of behavior with outcomes.

motive of the business within which a person works (this factor was also suggested by Victor and Cullen [1988] and Trevino [1986]). Personal normative beliefs represent the *moral obligation* a person feels to perform or to not perform a behavior, as suggested by Schwartz and Tessler (1972). Last, *awareness of consequences* has been proposed as a potential influence in decision-making (Rest, 1979; Eining and Christensen, 1991).

A functional representation of the proposed ethical/unethical decision model for this study is:

D = f (societal environment, belief system, personal values, personal environment, professional environment, legal environment, business environment, moral obligation, and awareness of consequences.) (1)

where: D= Decision [acceptable (ethical) versus unacceptable, unethical)]

An individual's judgment of what is ethical or unethical was measured using five IS-related ethical dilemmas described in scenarios (see Appendix for Case A through Case E). The specific scenarios used were chosen based on relevancy to IS ethical dilemmas. They represent a range of ethical issues involving copyright, piracy, privacy, and security. Each scenario describes an ethical situation and presents a response or decision made by the individual(s) in the case. Subjects were given the five scenarios and were asked to determine whether the behavior of the individual presented in the case was acceptable (ethical) or unacceptable (unethical). It should be noted that the words "acceptable" and "unacceptable" were used interchangeably with "ethical" and "unethical" during the administration of the instrument.

After judging the behavior in each scenario the respondents were asked to indicate the extent to which the list of factors described earlier had influenced their decision. The degree of influence could range from *no* influence to *great* influence using a 5-point scale. Some observations were not usable since some responses were incomplete.

Multiple discriminant analysis was used to determine the effect of each factor on the acceptable/unacceptable ethical decision. Discriminant analysis is the classification of an observation x, into one of several populations, each of which have density functions. The analysis involves deriving the linear combination of the independent variables that best discriminates between the *a priori* groups. This is achieved by maximizing the between group variance relative to the within group variance. The linear combinations for a discriminant analysis are derived from an equation of the form:

$$Z = W_1X_1 + W_2X_2 + W_3X_3 + \ldots + W_nX_n \quad (2)$$

where Z is the discriminant score, the W's are the discriminant weights, and the X's are the independent variables. Stepwise discriminant analysis was used to determine the significant factors that discriminate between decisions of what is acceptable or unacceptable. Based on the stepwise discriminant analysis, discriminant analysis was done with only significant variables allowed to remain in the function. The classification rates based on each scenario's significant factors (*reduced model*) and the classification rates based on the nine-factor model (*full model*) were compared.

RESULTS AND DISCUSSION

This study is part of ongoing research into the influential factors in ethical decision-making. The present study involved 307 subjects who were students in various business courses. Table 2 lists some of the demographic information about the subjects. The average age was 21.6 years, with a range of 17 to 49 years. The average GPA was 3.0 and the average years of work experience was 2.9. There were 128 female subjects and 179 male subjects. Every college level, from freshman to graduate, was represented but juniors (46%) were the largest group. Most of the subjects were single (215), with only 32 married and 5 divorced.

Table 3 presents a summary of the decisions and the degree of influence attributed to the nine factors theorized to influence the subjects' assessment of whether an action was acceptable or unacceptable. In four of the five cases the majority of subjects decided that the actions of the person described in the case were unacceptable (i.e., unethical). The frequencies for the degree of influence (from none to great) indicate the extent to which subjects said factors were influential in deciding about a case, as well as indicating which factors the subjects said were not pertinent.

Discriminant analysis was used to assess how well the significant factors (based on stepwise discriminant analysis) can classify the behavior described in each scenario as acceptable or unacceptable. The results of the discriminant analysis are shown in Table 4. The significant variables and the respective classification values are listed. In all five cases, *moral obligation* (personal normative beliefs) was a significant factor, based on stepwise discriminant analysis. This confirms the findings of Schwartz and Tessler (1972) that personal normative beliefs are a significant influence on behavior intention. For comparison, the classification rates for the full nine-factor model are also listed. Case A produced a successful classification rate of 77.6% based on only three of the original nine factors used in this study. Case B had a classification rate of 73.1% using four factors. The *moral obligation* factor was the only one found to be significant for Case B and the successful classification rate was 75.3% with that factor. The two significant factors for Case D produced a classification rate of 66.9%. The classification rate for Case E was 74.8% based on three factors. The comparison between the full model and the more parsimonious models show there is little difference in classification rates.

The results of this study give some insight into ethical decision-making and the factors that influence such decisions. For instance, Case A deals with

Table 2: Subject Demographics

Number of subjects	=	307*
Age:		
Average	=	21.6 years
Range	=	17 to 49
GPA:		
Average	=	3.0
Range	=	1.5 to 4.0
Work Experience:		
Average	=	2.9 years
Range	=	0 to 28
Sex:		
Females	=	128
Males	=	179
College Level:		
Freshman	=	32
Sophomore	=	70
Junior	=	141
Senior	=	58
Graduate	=	6
Marital Status:		
Married	=	32
Single	=	215
Divorced	=	5

* N varies for some analyses because of incomplete responses.

a programmer who manipulated a bank's accounting system to hide his overdrawn account. Stepwise discriminant analysis found *moral obligation, awareness of the consequences, legal environment*, and *belief system* to be significant factors in determining acceptable/unacceptable behavior. From Table 3, the frequencies show that 199 subjects (76% of the subjects who said the behavior was unacceptable) felt they would have a strong moral obligation to take corrective action.

In Case A the *legal* factor was also significant. The majority of those who said this was unacceptable (166 out of 262) indicated that the *legal* factor was very influential in their decision. The majority of those who said this was acceptable (29 out of 42) indicated that the *legal* factor wasn't influential in their decision. In fact, in discussions about this case with students (after the

Table 3: Factors and Weights used for Ethical Decisions

Frequencies	**Case A** (Programmer manipulates accounting system.)		**Case B** (Software sent in error was kept.)		**Case C** (Company equipment used on personal time.)		**Case D** (Used program without paying requirered fee.)		**Case E** (Copied data made accessible during contract work.)	
	Acceptable	Unacceptable	Acceptable	Unacceptable	Acceptable	Unacceptable	Acceptable	Unacceptable	Acceptable	Unacceptable
Decision:	42	262	121	184	265	39	97	205	120	182
Variable										
Societal environment										
none	9	40	23	25	59	5	19	22	14	14
little	15	65	26	61	74	12	27	55	35	29
moderate	16	101	34	63	79	16	32	82	35	53
much	1	44	31	28	41	6	15	36	30	56
great	1	13	8	8	13	-	5	12	7	30
Belief system										
none	6	26	18	9	50	6	12	17	14	10
little	12	39	29	32	58	5	30	26	23	23
moderate	11	50	38	36	82	10	25	64	46	43
much	11	84	27	48	52	13	23	63	24	64
great	-	64	10	60	25	5	8	38	15	42
Personal values										
none	2	6	6	1	23	2	6	6	6	2
little	4	8	18	11	39	3	8	24	16	11
moderate	11	39	41	29	73	9	32	52	42	40
much	12	102	34	65	84	18	35	71	34	67
great	12	109	22	79	45	7	16	54	23	62
Personal environment										
none	7	19	15	5	37	4	10	10	11	8
little	8	31	22	26	51	10	16	40	23	26
moderate	16	86	35	58	79	12	39	63	39	57
much	3	72	41	56	68	11	23	63	31	56
great	7	52	9	38	30	2	10	31	17	35

Table 3: Factors and Weights used for Ethical Decisions (continued)

Professional environment										
none	5	10	24	24	27	3	12	10	7	10
little	7	30	43	49	44	5	22	28	15	12
moderate	10	63	32	51	71	12	33	56	39	52
much	13	84	17	41	86	10	18	71	42	51
great	7	77	5	17	37	9	13	42	19	57
Legal environment										
none	6	13	27	22	58	2	14	8	15	10
little	14	33	41	33	76	11	23	25	18	14
moderate	9	52	26	59	68	15	30	55	30	32
much	8	75	17	36	43	5	18	69	33	48
great	5	91	11	34	20	6	12	49	25	78
Business environment										
none	5	18	30	26	37	2	18	11	15	13
little	8	38	35	47	47	6	18	28	21	24
moderate	13	65	29	58	71	7	26	64	33	50
much	10	77	18	33	69	11	26	57	34	45
great	6	67	10	20	41	12	10	48	19	50
The following two items were scored on a 5-point scale with two anchors:										
Moral obligation										
no obligation	7	3	48	23	121	1	21	14	20	4
.	10	9	40	27	78	8	28	18	28	13
.	14	54	26	50	50	13	33	74	40	36
.	8	112	5	49	11	9	13	53	25	57
strong obligation	3	87	3	36	5	8	3	48	8	72
Awareness of consequences*										
should have	3	12	12	6	14	-	6	7	5	2
.	4	9	14	11	20	3	11	10	9	6
.	10	15	32	20	56	6	23	32	28	12
.	12	31	21	38	62	10	24	44	21	26
should not have	13	198	43	110	115	20	34	115	59	136

A dash (-) represents a zero weight.
Variables whose frequencies are in bold were determined to be significant variables based on stepwise discriminant analysis.
* The actual wording regarding awareness of consequences varied based on the scenario.

Table 4: Classification Rates for Factors Used

	Predicted	Model Classification Rate Unacceptable	Acceptable	Total	Reduced (Significant)	Full
Case A*						
moral obligation	Unacceptable	206	56	262		
awareness of consequences	Acceptable	12	30	42		
belief systems	Total	218	86	304	77.6%	79.1%
legal environment						
Case B*						
moral obligation	Unacceptable	131	53	184		
personal values	Acceptable	29	92	121		
societal environment	Total	160	145	305	73.1%	72.5%
legal environment						
awareness of consequences						
Case C*						
moral obligation	Unacceptable	30	9	39		
	Acceptable	66	199	265		
	Total	96	208	304	75.3%	78.3%
Case D*						
moral obligation	Unacceptable	132	73	205		
legal environment	Acceptable	27	70	97		
	Total	159	143	302	66.9%	70.3%
Case E*						
moral obligation	Unacceptable	137	45	182		
legal environment	Acceptable	31	89	120		
awareness of consequences	Total	168	134	302	74.8%	79.1%

* Factors listed were significant variables in the reduced model.

ethics survey had been given) some indicated they would do the same things as the person described in the scenario, particularly if they were confident they would not get caught. Given these results about Case A, business managers may be able to influence the individual's decision to misuse computer technology by making business policies clear to employees. Such policies should delineate what is legal and acceptable behavior and what is unacceptable and punishable behavior. Business could also emphasize that detection and monitoring techniques are used.

In Case B a person decided to keep a software program that was sent in error from a mail-order store. For this case, *moral obligation* was a significant influence, as well as *personal values, societal environment, legal environment*, and *awareness of consequences*. Case C described a situation where a programmer used company equipment to write programs for his friends but he did so on his own time on the weekend. This is the only case where the majority of subjects (87%) considered the behavior of the person in the scenario acceptable. Again, *moral obligation* was significant in predicting whether subjects would consider the behavior acceptable or not. One hundred ninety-nine subjects who judged the behavior as acceptable, felt little or no moral obligation to take any action in this case. Case D presented a situation in which a person was inadvertently given free access to a program that normally required a fee and which was considered proprietary. The person opted to use the program without paying the usage fee to the service company. For this case the *moral obligation* and the *legal environment* were significant factors in determining whether the behavior was acceptable or not.

The final case involved a marketing company employee that did some data processing on contract for a government agency. The agency's data contained information about children and their parents. At the request of the employee's boss, the employee made a copy of the data for the company's use after determining that the contract with the government agency contained no divulgement restrictions. Again, *moral obligation* and the legal environment were significant, as well as *awareness of consequences*.

Reviewing significant factors across cases, we find *moral obligation* to be the only factor that is significant in all cases. Yet, legal environment and awareness of consequences are significant in three of the cases. Given these results, the sense of an obligation to take corrective action was a significant indicator of a person's judgement as to whether the action was acceptable or unacceptable. Students who felt an obligation to take corrective action usually felt the action in the case was unacceptable. Subsequent discussions with students revealed that knowledge of the negative consequences of an action could affect what a person would do. Awareness of the consequences could alter how people behave. Similarly, the knowledge of the legality of the event is significant. Students revealed that knowledge of the specific laws and terms of a contract could influence their decision.

Businesses could affect the ethical behavior of the employees by using some of the results presented here. For example, a clear policy statement with expectations and consequences appears to be a factor that could make a difference. Employee's knowledge of what behavior is acceptable and what behavior is unacceptable, along with its consequences, could be effective in deterring some types of computer misuse. Given the results of this research, businesses could develop a plan to educate employees concerning company policies and codes of conduct. However, some significant factors, such as moral obligation (personal normative beliefs) may not be easily altered through education. Managers should also consider detective measures if they are concerned with unacceptable behavior.

CONCLUSIONS

Ethical decision-making is theorized to be affected by several factors. The proposed model suggests that ethical decisions are influenced by religious values or beliefs, societal or cultural values, personal values, normative beliefs, awareness of the consequences of behavior and the environments within which we live and work — personal, professional, legal, and business. The purpose of this research was to determine whether factors can be identified which influence the assessment of behavior as ethical or unethical. Based on the results of this research, factors which influence the ethical decision can be identified. As reported in Table 4, the factors that influence our judgment of ethical or unethical behavior vary by case.

In a practical sense, this only provides a starting point for understanding what influences people's ethical decision-making. This research assists management in understanding what the influential factors are and which of these managers could use to guide employees and reduce the misuse of computer technology. With training programs, management examples, the formulation of codes of conduct, and the enforcement of company policies and rules, companies may be able to deter some computer misuse.

It was noted at the beginning of this paper that this study is part of an ongoing research effort into influential factors and ethical decision-making. Additional work is underway. For instance, recent research by Loch and Conger (1996) found differences between men and women in their ethical decision-making and what affected their decision. Further study into such differences between men and women, as well as other characteristics of the subjects such as age and work experience, could be beneficial in explaining an individual's ethical judgment and ethical intentions. Also, the next phase of this study incorporates concepts from Robin et al. (1996) and Jones (1991) about the situation-specific aspect of ethical decision-making and the perceived importance of a particular ethical issue.

APPENDIX: CASES FOR THE STUDY

Case A

A programmer at a bank realized that he had accidentally overdrawn his checking account. He made a small adjustment in the bank's accounting system so that his account would not have an additional service charge assessed. As soon as he made a deposit that made his balance positive again, he corrected the bank's accounting system. Was the programmer's modification of the accounting system acceptable, questionable, or unacceptable?

Case B

A computer user called a mail-order computer program store to order a particular accounting system. When he received his order, he found out that the store had accidentally sent him a very expensive word processing program as well as the accounting package that he had ordered. He looked at the invoice, and it indicated only that the accounting package had been sent. The user decided to keep the word processing package. Was the user's decision to keep the word processing package acceptable, questionable, or unacceptable?

Case C

A computer programmer enjoyed building small computer systems to give his friends. He would frequently go to his office on Saturday when no one was working and use his employer's computer to develop systems. He did not hide the fact that he was going into the building; he had to sign a register at a security desk each time he entered. *Was the programmer's use of the company computer acceptable, questionable, or unacceptable?*

Case D

A commercial time-sharing service offered use of a program at a premium charge, the program to be used only in the service company's computer. A user obtained a copy of the program accidentally, when the service company inadvertently revealed it to him in discussions through the system (terminal to terminal) concerning a possible program bug. All copies of the program outside of the computer system were marked as trade secret, proprietary to the service, but the copy the customer obtained from the computer was not. He used the copy of the program after he obtained it, without paying the usage fee to the service. *Were the actions acceptable , questionable, or unacceptable?*

Case E

A marketing company's employee was doing piece work production data runs on company computers after hours under contract for a state government. Her moonlighting activity was performed with the knowledge and approval of her employer. The data were questionnaire answers of 14,000 public school children. The questionnaire contained highly specific questions on domestic life of the children and their parents. The government's purpose was to develop statistics for behavioral profiles, for use in public assistance programs. The data included the respondents' names, addresses, and so forth. The employee's contract contained no divulgement restrictions, except a provision that statistical compilations and analyses were the property of the government. The employer discovered the exact nature of the information in the tapes and its value in business services his company supplied. He requested that the data be copied for subsequent use in the business. The employee decided the request did not violate the terms of the contract, and she complied. *Were the actions acceptable, questionable, or unacceptable?* (Adapted from Dejoie, Fowler, and Paradice, Ethical Issues in Information Systems, (Boyd & Fraser, 1991).

ENDNOTES

[1] Ethical is defined as in accordance with *accepted* standards of behavior (American Heritage Dictionary, 1985).

REFERENCES

Aiken, R.M. (1988). "Reflections on Teaching Computer Ethics," Paper presented at the Fourteenth SIGCSE Technical Symposium on Computer Science Education, 8-11.

Ajzen, I. & Fishbein, M. (1980). *Understanding Attitudes and Predicting Social Behavior,* Prentice-Hall, Engelwood Cliffs, NJ.

Anderson, R. E., Johnson, D. G., Gatterbarn, D. & Perrolle, J. (1993). "Using the New ACM Code of Ethics in Decision Making," *Communications of the ACM*, 36(2), 98-107.

Beck, L. and Ajzen, I. (1991) "Predicting Dishonest Actions Using the Theory of Planned Behavior," *Journal of Research in Personality,* 25, 285-301.

Bommer, M., Gratto, C., Gravander, J. & Tuttle, M. (1987). "A Behavioral Model of Ethical and Unethical Decision Making," *Journal of Business Ethics,* 6 (May), 265-280.

Conner, K.R., & Rumelt, R.P. (1991). "Software Piracy: An Analysis of Protection Strategies," *Management Science,* 37(2), 125-139.

Cougar, J.D. (1984). "Providing Norms on Ethics to Entering Employees," *Journal of Systems Management*, 35(2), 40-43.

Cougar, J.D. (1989). "Preparing IS Students to Deal with Ethical Issues," *MIS Quarterly,* 13(2), 211-218.

Dejoie, R., Fowler, G. & Paradice, D. (1991). *Ethical Issues in Information Systems,* Boyd & Fraser.

Eining, M.M. & Christensen, A.L. (1991). "A Psycho-Social Model of Software Piracy: The Development and Test of a Model," in *Ethical Issues in Information Systems*, R. Dejoie, G. Fowler, and D. Paradice (eds.), Boyd and Fraser, Boston, MA.

Ferrell, O.C. & Gresham, L.G. (1985). "A Contingency Framework for Understanding Ethical Decision Making in Marketing," *Journal of Marketing,* 49(Summer), 87-96.

Heide, D. & Hightower, J.K. (1988). "Teaching Ethics in the Information Sciences Curricula," *Proceedings of the 1988 Annual Meeting of the Decision Sciences Institute,* 1, 198-200.

Jones, T.M. (1991) "Ethical Decision Making by Individuals in Organizational; An Issue-Contingent Model," *Academy of Management Review,* 16(2), 366-395.

Kohlberg, L. (1969). "Stage and Sequence: The Cognitive-developmental Approach to socialization," in *Handbook of Socialization Theory and Research,* D. Goslin (ed.), Rand McNally and Co., Chicago, IL.

Loch, K.D. & Conger, S. (1996). "Evaluating Ethical Decision Making and Computer Use," *Communications of the ACM* 39(7), 74-83.

Oz, E. (1990). "The Attitude of Managers-To-Be Toward Software Piracy," *OR/MS Today* 17(4), 24-26.

Paradice, D.B. (1990). "Ethical Attitudes of Entry-Level MIS Personnel," *Information and Management,* 18, 143-151.

Parker, D.B. (1980). *Ethical Conflicts in Computer Science and Technology,* AFIPS Press, Reston, VA.

Research Institute of America. (1983)."Safeguarding Your Business Against Theft and Vandalism," *Computer Crime Digest,* 5(November).

Rest, J.R. (1979). *Development in Judging Moral Issues,* University of Minnesota Press, Minneapolis, MN.

Robin, D.P., Reidenbach, R.E. and Forrest, P.J. (1996) "The Perceived Importance of an Ethical Issue as an Influence on the Ethical Decision-making of Ad Manager," *Journal of Business Research,* 35, 17-28.

Saari, J. (1987). "Computer Crime - Numbers Lie," Computers and Security 6(June), 111-117.

Schwartz, S.H. & Tessler, R.C. (1972). "A Test of a Model for Reducing Measured Attitude-Behavior Discrepancies," *Journal of Personality and Social Psychology*, 24(2), 225-236.

Simon, H. (1960). *The New Science of Management Decisions,* Harper and Row, New York, NY.

Straub, D.W. (1986). "Deterring Computer Abuse: The Effectiveness of Deterrent Countermeasures in the Computer Security Environment," Doctoral Dissertation, Indiana University School of Business, Bloomington, IN.

Straub, D.W. & Nance, W.D. (1990). "Discovering and Disciplining Computer Abuse in Organizations: A Field Study," *MIS Quarterly* 14(1), 45-60.

Trevino, L.K. (1986). "Ethical Decision Making in Organizations: A Person-Situation Interactionist Model," *Academy of Management Review* 11(3), 601-617.

Victor, B., & Cullen, J.B. (1988). "The Organizational Bases of Ethical Work Climate," *Administrative Science Quarterly*, 101-125.

Zalud, B. (1984). "IBM Chief Urges DP Education, Social Responsibility," *Data Management* 22(9), 30, 74.

Chapter 6

Ten Lessons that Internet Auction Markets Can Learn from Securities Market Automation

J. Christopher Westland
Hong Kong University of Science and Technology

Internet auction markets offer customers a compelling new model for price discovery. This model places much more power in the hands of the consumer than a retail model that assumes price taking, while giving consumers choice of vendor and product. Models of auction market automation has been evolving for some time. Securities markets in most countries over the past decade have invested significantly in automating various components with database and communications technologies. This paper explores the automation of three emerging market exchanges (The Commercial Exchange of Santiago, The Moscow Central Stock Exchange, and Shanghai's Stock Exchange (with the intention of drawing parallels between new Internet models of retailing and the older proprietary networked markets for financial securities.

INTRODUCTION

Internet auction markets, such as those offered by Amazon, eBay, Priceline, OnSale, and CNET's Shopper.com, are earning increased business and investment by offering customers a new model for price discovery. This model places much more power in the hands of the consumer than a retail model that assumes price taking, while given consumers choice of vendor and product. The monetary impact on retailing is currently small (Internet E-commerce in the U.S. in 1998 totaled US$7.8 billion. Compare this to Wal-

Previously Published in the *Journal of Global Information Management, vol.8, no.1*, Copyright © 2000, Idea Group Publishing.

Mart's retail sales in 1998 of $130 billion, or total U.S. retail sales in 1998 of $1.7 trillion. But Internet sales are growing faster than traditional retailing, reflecting the appearance of auction based price discovery that has been made possible by Internet automation of many retailing outlets.

Models of auction market automation have been evolving for some time. For example, securities markets in most countries over the past decade have invested significantly in automating various components with database and communications technologies. Various other technologies have a mechanized securities markets for over a century; for example, stock tickers have provided automated real-time reporting of securities prices for nearly a century. Without automation, markets are constrained to operate at the speed of their human facilitators — frequently too slow and localized for complex or high volume market services.

This paper explores the automation of three emerging market exchanges (The Commercial Exchange of Santiago, The Moscow Central Stock Exchange, and Shanghai's Stock Exchange) with the intention of drawing parallels between new Internet models of retailing and the established proprietary networked markets for financial securities.

Emerging market innovations can elucidate more clearly specific issues arising in automation, because the projects are not hampered by tradition, volumes may be smaller, trading more localized, and offerings more homogeneous. It may thus be easier to discern the rationale behind a particular technology choice.

Each of these three exchanges has, over the past decade, experimented with information technology appropriate for its market. Each discovered unique issues and pitfalls in automating its particular exchange operations. This article summarizes what was learned from market automation. The implications for automation of retail markets in general (what has come to be called electronic commerce, or *e*-commerce — are drawn from the lessons learned in automating these securities markets.

CASE STUDY IN THE AUTOMATION OF SHANGHAI'S STOCK EXCHANGE

Deng Xiaoping initiated the "responsibility system" in 1979 – a capitalist innovation which abolished central quotas and allowed farmers and some township enterprises to sell their goods on the open market. Many became rich in the ensuing decade. Savings grew with the growing wealth of the populace, endowing China with one of the highest savings rates in the world — between 35% and 40% of GDP over the past decade. With the introduction of economic reforms in the 1980s, average annual growth was pushed to nearly

10%. This modernization required substantial amounts of investment capital, which was the main role of the Shanghai Stock Exchange, China's main securities market.

Most trading takes place in what are called "A" shares, which may be traded on the floor, but more often, trading bypasses the floor completely, and is handled through automated systems. The Shanghai Stock Exchange issues a security card that identifies an individual as being authorized to trade in listed securities. The security card allows traders to bypass brokerage firms (and reduce brokers' commissions). This provides a more efficient use of the floor, by channeling routine trades directly to computer matching. Trades are captured at a brokerage room when traders post an order with the clerk. The order is posted in the Exchange's system through computer-to-computer communication. Matching takes place automatically. Security card holders still have to go through a brokerage firm to buy or sell securities and to settle and clear the transaction. Security cards make the job of brokerage firm easier by identifying the client to the brokerage, and allowing automated management of the client's account balance. In contrast, "B" shares (which are designed for trading by overseas investors) are only traded through brokers and on the floor.

Liquidity is quite good in "A" shares, where there is widespread participation by China's populace through approximately 3000 brokerage rooms around China. Turnover is around five times per year. "B" shares are less liquid, and there has been considerable interest in engaging more overseas investors in their purchase. Much of the current trading in "B" shares is done by Chinese within China, though they must pay for these shares with dollars in accounts outside China.

Table 1: Market Capitalisation of Selected Markets in 1997

Exchange	Capitalisation Billions of US$[1]	Stockmarket Turnover % of Market Capitalisation[2]
New York	7600	84
Tokyo	2685	47
London	1500	42
Hong Kong	440	46
China (Shanghai-Shenzhen)	140	225
Russia	100	10

[1] Asiaweek, Feb. 28th 1997, p. 62
[2] The Economist, June 14th 1997, p.124

"A" shares can be traded by any trader with a Shanghai Stock Exchange issued security card. These traders can trade directly with the Exchange's computer system. Thus orders may come directly from trading counters — where there is a transaction-by-transaction cash settlement and exchange of securities — or through member brokers — where the Exchange provides net settlement of brokers accounts. Settlement is completed in the same day as the trade (T+0). Exchange commissions are approximately 0.65% of trades.

As in many emerging markets, price volatility can be more strongly influenced by the money coming into the market than by the business fundamentals of the traded firm. Volatility in the Shanghai market is very high by the standards of most developed economies, a situation that poses difficulties to orderly trading. This tends to compound the Exchange's challenges in the face of problems inherent in securities laws that are still evolving, and, in the difficulties faced by the China Securities Regulatory Commission in policing insider trading, misuse of funds, and false disclosure. Volatility is tightly monitored — important in a market where there is a rapidly expanding group of investors of varying levels of sophistication. The Exchange will not allow prices for a given security to vary more than 10% from the prior day's closing price. When the 10% mark is reached, trading is not halted. Rather the settlement prices simply are not allowed to exceed 10% of the prior day's price. This limit does not apply to initial public offerings (IPOs) on the first day; it only takes effect on the second day after issue.

The Shanghai Stock Exchange (and its sister exchange in Shenzhen) use a continuous double auction, order-driven trading system, assisted by a computer network to transfer order information from brokers to the floor, and back again. During opening hours, information on trade prices and volumes is continuously disseminated to traders, and buy and sell orders are continuously received. Both have supporting clearing houses for transaction completion and settlement.

The Exchange floor opens Monday through Friday from 9:30 AM until 11:30 AM (at which time prices are frozen), and re-opens after lunch from 1:00PM to 3:00PM. Opening prices are generally set to clear the maximum number of outstanding bid and ask orders on hand at 9:15 AM. They may be set differently if there are substantial imbalances between buying and selling, and it is felt that a different price is appropriate to clear the market. The opening prices for thinly traded stocks will be set at their closing prices from the last business day.

The trading system runs on a Hewlett-Packard HP9000-T500 computer, with peak processing of 5000 transactions per second. This is sufficient to handle peak Exchange volume of two million transactions per day, with peak periods at the beginning and end of the trading day. The HP9000 is, with the addition of processing boards, scalable to 20 million transactions per day. Cisco routers regulate traffic on the local and wide area networks controlling traffic to and from brokerage rooms throughout China. These rely on a fault tolerant combination of direct digital network, satellite and analog transmission. The Exchange has been clever in applying a cost conscious set of components for exchange automation that make effective use of both microcomputer and state-of -the-art computer and communications technology.

Trading on the Shanghai Stock Exchange is dominated by small investors – 99.4% of the Shanghai Stock Exchange's clients are individuals; another 0.6% are institutions with only a vestige of direct participation by securities houses. The Shanghai Exchange has implemented a number of innovations to bring online trading to the small investor. China's limited telecommunications capacity requires several levels of support to reach its desired customer base. Brokerage rooms provide the equivalent of Internet cafés for those who do not have a telephone, modem and computer at home. In addition, Real-time stock transaction quotations are provided online, free to subscribers of Shanghai's cable television service. A computer card can provide the same cable TV quotation service, but on a computer screen; over 50,000 were sold in their first year. Traders with a modem and computer can open a trading account with one of the local banks. The Exchange then provides them free connectivity software and free real-time quotations, allowing trading in and out of that bank account. The StockStar electronic trading Internet site provides services similar to E*Trade in the U.S., though the level of support provided to traders directly from the Exchange has provided it little latitude for profit.

Information dissemination on exchange transactions is a significant problem because of China's size. China is a continent sized country which needs sophisticated communications systems to allow access to its exchanges. Shanghai's Stock Exchange has invested in two networks: STAQ, a national-wide system for broadcasting transaction prices; and NET, an automatic security trading system for stocks as well as government bonds. Sixteen networked securities trading centers with broadcast and screen facilities have been established around China. In 1993 these received new fiber optics and satellite communication systems which replaced dedicated telephone lines. Communications are now supported by an optical fiber network throughout the city of Shanghai. A combination of local fiber loops, dedicated satellite communication systems (using the AsiaOne satellite for communications linkage) and telephone lines offers two-way communication with 3000 trading counters in over 300 cities around China. There are roughly five million investors around China, and around half of the monetary trading volume originates outside Shanghai.

The Exchange floor has gone through substantial changes in its first decade. It has supported as many as 6000 seats (i.e. booths with microcomputers and telephones) spread over eight trading halls, representing the 500+ authorized financial institutions who are members of the Exchange. Firms must have registered capital of more than RMB 5 million to be members. The new 27-floor Exchange building in the PuDong (literally: Shanghai East) area of Shanghai will dedicate all space from the ground up to the ninth floor to Exchange operations. The exchange floor in PuDong holds 1700 traders on a floor double the size of Tokyo's and triple the size of Hong Kong's floor. Most business though is conducted off the floor, through terminals and automated matching, and floor brokers spend a disproportionate time reading their newspapers. The floor provides no formal trading or market-making functions. It provides a focal point for market management and supervision, and for market sentiment. Li Quian, an exchange official, explains that "It is also a symbol of Shanghai's success."

To facilitate transparency of market activities, the communications network automatically disseminates trading information:

1. to the Exchange's trading room display (a large digital screen on the trading floor),
2. to a telephone inquiry network for people not on the trading floor,
3. to over 20 news organizations, TV and radio stations,
4. to Reuters, Telerate, and other global financial services, and
5. publishes a newspaper, *Shanghai Security*, with a circulation of several hundred thousand.

There are 1000 minor satellite downlink stations, and one major uplink station in PuDong. In brokerage rooms in Shanghai and other major cities in China, satellite systems typically provide backup order placement functionality in case telephone lines are not operable. In remote parts of China, such satellite systems are the sole communication channel allowing investors to participate in the trading of securities listed on the Shanghai Stock Exchange. Since 1993, the State has encouraged individuals and institutions outside Shanghai to participate in securities listed on the Shanghai Stock Exchange, which it would like to truly be a national Exchange. There are currently more transactions posted by traders outside than inside Shanghai.

The Exchange actually engages two systems vendors to provide software and services for the collection and dissemination of trading information. They provide basically the same service. Since customers can choose either one, this provides considerable incentive for both vendors to maintain their quality of service and sophistication of software.

China's Securities and Exchange Commission worked to insure reliability of disseminated information. It has installed stringent controls over the information reported by companies, and has prohibited the accounting profession from certifying forecasts of corporate performance.

Information on all transactions is also transferred to the Exchange's market monitoring group which attempts to control insider trading, rumors, and collective efforts to control prices. First priority in automation of exchange functions has gone to the "A" shares. "B" shares are purchased in dollars, though denominated in renminbi (RMB), and information on those shares is broadcast in both English and Putonghau. Thus additional systems development is required in supporting "B" shares.

Customer participation in any auction market requires that customers have confidence that orders will be priced fairly, and not subject to manipulations which would put them at a disadvantage. Chinese securities laws impose some order on the off-floor brokerage function. Brokers are required to comply with customers' explicit terms for transactions conducted on their behalf, including: (1) type of security, (2) volume of trade, (3) bidding conditions and margin level, and (4) time of authorization. To prevent "insider" trading, employees of securities authorities, managers of the securities exchanges, employees of the broker handling the transaction and employees of the governmental agency regulating or controlling the issuing company are prohibited from trading in its shares (Zhao and Li, 1992). Compliance with these guidelines is enforced by a Monitoring Department and a Compliance Department (i.e., investigation) in the Exchange organization.

Seats in the Shanghai Stock Exchange are classified as "real" and "virtual." Virtual seats are either remote terminals or handsets that are connected to the matching software through DDN, satellite uplink, or analog telephone link. There exist 1000 low bandwidth satellite uplinks for posting orders remotely, and one broadband satellite uplink (in Pudong at the Shanghai Stock Exchange) to disseminate bid, ask and settled trade prices and volumes.

Order placement security is enabled by either a touch tone or a magnetic card verification system. The touch tone system requires the cashier/broker to enter trader identification, security code, password, limit price and quantity of the buy or sell order. The magnetic card system (i.e., a standard sized credit card for stock transactions) enters trader identification and security code automatically requiring keying only of password, limit price and quantity of the buy or sell order.

Magnetic card system is provided by the Shanghai Stock Exchange and provides an additional level of security over trading. Individuals wishing to trade in the Shanghai Stock Exchange need to procure a magnetic security card from the Exchange. This card uniquely identifies the trader's account with the Exchange. Since the market is a cash market, every share is owned by someone, and this is recorded in the Exchanges databases. The Exchange handles its own transfer/clearing accounting. Through this account identification, traders can obtain information about their account position from the Exchange.

Banks and brokerage houses issue separate debit cards which allow traders to buy and sell without exchanging cash. These reflect account balances with the bank. Traders who do not have a security card can trade, with the broker acting on their behalf. Some banks incorporate the security card number and information into the debit card. Not all brokers have a direct connection to the Exchange. Particularly in remote areas of China, brokers need to dial into their seat on the floor (or through satellite uplink) to place orders. Even in Shanghai, some brokerage firms insist on their own staff dialing for traders, even if the traders can directly place orders with the exchange. This provides an additional layer of security and control over trading.

Clearing and settlement are the responsibilities of a wholly-owned subsidiary of the Shanghai Stock Exchange — the Shanghai Securities Central Clearing and Registration Corporation. Central depository, trades and clearing are all paperless. There is no need to print a physical copy of the security, as a database is maintained of ownership of all shares.

A particularly exciting innovation in Chinese securities trading in recent years has been the introduction of neighborhood brokerage rooms to encourage the investment of funds by China's people, who have one of the highest savings rates in the world. The immense popularity of trading counters and brokerage rooms in China has done much to abate liquidity shortages that dogged the market until recently. Before the widespread installation of trading rooms, China's securities markets suffered from a lack of capacity to absorb large buying or selling pressure without causing severe adverse price movements.

China's investment banks rely on a clever mix of appropriate technologies, cost effectively delivered, to bring their services to the people. Brokerage rooms provide just the right amount of technology for a country with straggling telecommunications infrastructure, and few PCs. Electronic investment technology is brought close to every home, but not into it. Investors meet in a pleasant and convivial neighborhood atmosphere. In China, as elsewhere, more is not always better. Appropriate technology is intelligently embraced which fits the customer and the tasks at hand.

Small investors are served in a large hall crowned with an electronic bulletin board broadcasting the latest securities prices. In the front of the hall are windows through which clerks can process, buy or sell orders for listed securities. The bank also provides software for technical analysis of securities prices on computers at the front counter. Over 50 algorithms for technical analysis may be called up in the software; moving averages and other cycle analysis can be customized by the user.

Major investors – those with over RMB 500,000 in the market–are treated to their own desks and computers (complete with technical analysis software) in one of several rooms cloistered within the maze of halls leading from the main brokerage room. Despite the greater stakes, the unique and convivial atmosphere of investing still permeates these rooms. Though orders could be posted directly to the Stock Exchange's electronic matching system directly by these computers, the China Commercial Bank chooses to act as an intermediary. A secretary is provided for each of the VIP rooms, and actually places buy or sell orders at the request of the investors. Besides providing a professional touch for the customer-investor, this also provides the bank with an additional modicum of control.

CASE STUDY IN THE AUTOMATION OF MOSCOW'S CENTRAL STOCK EXCHANGE

Russian president Boris Yeltsin's government has rapidly moved to open markets since its economic reform program was launched in January 1992—by freeing prices, cutting defense spending, eliminating the old centralized distribution system, completing an ambitious voucher privatization program in 1994, establishing private financial institutions, and decentralizing foreign trade.

The Soviet economy was founded on heavy industry– on large-scale factories, smelters, refineries, extraction and processing of raw materials and natural resources. Much of Soviet budgetary allocation through the 1950s and 1960s was directed toward infrastructure development. The country formed a national electricity grid (now a joint stock company known as "Unified Energy Systems"), a gas exploration, drilling and refining behemoth (now joint stock company "Gazprom"), a nationwide telecommunications system ("Svyazinvest" and "Rostelecom"), oil exploration and distribution corporations ("Lukoil" and "Surgutneftegaz," inter alia), car and truck manufacturers ("Logovaz," "Zil," and "Kamaz"), diamond and gold processing organizations. These industries provided the securities marketed on the Moscow Central Stock Exchange.

The establishment of modern financial markets and the modernization of banking helped the non-State sector to contribute approximately 75% of official GDP by 1997, up from 62% in 1994. This contribution may understate the true contributions from privatization. By some estimates, the official GDP figure of around US$ 700 billion (at purchasing power parity, with a bit over 10% contributed by exports) is only 50% of the actual GDP. The underreporting of economic activity to avoid an inequitable tax system obscures actual income. Russians save 32% of their income (slightly short of the Chinese savings rate) while the Russian government consumes only 15% of the official GDP (compared to 30% of GDP in the U.S.). These amounts are comparable to State sector consumption in China, where official figures understate GDP, and faulty tax collection is again to blame.

To privatize these firms, the Russian government issued vouchers, worth ten thousand rubles each, to every Russian citizen (children included). These could be picked up at local offices of the state bank for a nominal transaction fee; eventually some 144 million out of 147 million Russians received their vouchers. They could be exchanged for shares in companies through the mechanism of auctions (essentially they were currency, but could only be used to bid for shares. Markets grew for buying and selling vouchers; they were even sold from farmers kiosks just like carrots or cabbages.

Voucher funds' ability to influence enterprise performance depended on a number of factors, including management compliance with new policy, management disclosure practice, and support by local and regional government officials. With the absence of access to enterprises' financial information, however, voucher funds were severely restricted in their evaluative capacity–a significant liability for an instrument which was initially motivated by a need to value enterprise assets. Given voucher funds' relatively small ownership stakes, the funds also often required the cooperation of other shareholders in order to put forward new policies.

The Moscow Central Stock Exchange (MSE) opened in 1992 as the first over-the-counter marketplace in Russia. It became the focus of the fledgling equities trading market at the first stage of privatization. Russia's first standardised futures contract was designed by Moscow Stock Exchange's Clearing House in March 1994 to facilitate trading on USD/RUR exchange rate. In November 1995 a government bond (GKO) market evolved, and the MSE offered futures contracts on GKOs (index-linked, yield curve, when issued series, physically deliverable, etc.) with American-style options on futures as underlying. Later, contracts on Federal floating rate Notes (OFZs) were added to the list, making the Russian Bond futures market as an MSE franchise. Since then, the GKO and OFZ futures contracts have become increasingly important additions to the domestic treasury bond cash market. In June 1997, the Exchange's daily traded volume was between 30-40% of GKO and OFZ overall spot market turnover. These figures were considered proof of the growing need of large institutions to counterbalance their volatile GKO/OFZ holdings (borne out in the collapse of the debt market in August 1998).

The Exchange operates a screen-based multifunctional trading system. Trading is order-driven, in a continuous double-price auction, where orders may be placed by interactive remote access. The Exchange has implemented on-line clearing which has added liquidity and traded volume. The Exchange has engaged IBM to migrate from existing PC-based network to IBM RS/6000 servers. The IBM Global Network (IGN), utilised by the Exchange, makes viable the full-scale remote access and order routing for overseas brokers.

During the early days of equity trading, Russia's market consisted of brokers and dealers with stock shares, vouchers, and other privatization certificates. Price search was conducted over the telephone and fax (reminiscent of the over-the-counter market that preceded the US's National Association of Securities Dealers Automated Quotations system) and by traveling to distant regions by plane, train, and sometimes even on foot.

In 1994 and 1995, a broker's trip to a distant region where shares of Lukoil or Rostelecom were being sold at below-market prices by local residents (and, occasionally, by company officials) could net a 100% return on purchased shares merely from transporting them to Moscow. This may mark one of history's extreme cases in a liquidity-driven market. Improved market efficiency from automation dropped that geographic-liquidity-induced return to around 20% by 1997. As equity trading became more regulated, and information spread to the regions through paper and Internet information bulletins from 1995 on, brokers and traders moved to computerized trading systems such as the Russian Trading System and AK&M (similar to the NASDAQ).

Despite the large size of Russia's economy, total capitalisation of shares listed on the Moscow Stock Exchange still amounted to only around US$ 100 billion in 1997 (up from around $20 million two years earlier). Though Russia's market capitalisation is roughly that of China's stock markets, turnover was less than one-twentieth of China's (Table 1) at the time. However, the August 17th 1998 moratorium on repayment of Russian government debt threw the stockmarket into turmoil. By the end of September 1998, trades are less than $10 million per day (down from 10 times that amount only six month's earlier), and liquidity is almost non-existent as reflected in bid-ask spreads running from several 10s to several 100s of percent.

By 1997, a host of computerized systems existed in Russia's investment houses to actively and effectively process trades. While no singular system dominated the Russian market, larger investment houses were focusing on system implementation as trading volumes doubled and tripled. At Renaissance Capital, Russia's premier investment bank (headed by former CSFB Russia Director, Boris Jordan), for example, ten Sun Microsystems servers running proprietary software processed trades and analyzed positions for the company. Dealer boards—electronic devices that allowed equity traders to hold several telephone conversations at once —were installed in early-1997. Renaissance had a telephone system of 450 lines for its equity traders, with a Western-style trading floor with raised steel floors overlying high-bandwidth communications lines.

Most Western brokerage and trading houses import their own systems and procedures from the US or Europe. In contrast, Russian banks prefer to keep their proprietary information technology closed to outsiders. Russia has a well-educated workforce that excels in technology products such as software. By keeping systems and data feeds proprietary, banks can keep outsiders from knowing secrets that they believe provide them with a competitive edge.

Financial statements of listed firms typically are stated under both Russian Accounting Standards and International Accounting Standards (similar to US Generally Accepted Accounting Principles). Russian Accounting Standards provide more accurate accounting under inflation, or when asset valuation is ambiguous, as it is in an economy that still settles 40% of its industrial purchases through barter. Both Russian and Western banks use Russian accounting software packages, the two major packages provided by Diasoft and Program Bank. As the market matures, information technology providers should see considerable growth in business volume.

In contrast to more developed markets where settlement and clearing are instantaneous processes linked to price search, Russian back-office software "has to deal with a market in which completing a trade [often requires] sending a courier to Surgut to get the trade recorded in the company register." Furthermore, equity traders require at least one other employee in the back office to register trades, obtain stamps and seals, and comply with mountains of government regulations.

Russia's market regulating authorities fully understand and acknowledge the difficulties trading and brokerage houses face in a market dominated by telephone trade. According to Skate Press and several Moscow brokerages, at least 60% of trades are completed over the telephone and not registered with regulatory bodies. Furthermore, much of this trade takes place 'offshore', making trade tracking all the more difficult. This creates a major problem for both market makers and investors, as actual trading (market) prices often differ significantly from published (recorded) prices. As an example, for most of early-August 1997, recorded trade volume for Gazprom averaged 100,000 shares traded per day. Actual trade volume, according to several Moscow-based brokers, was at least twenty times that amount. Because settlement of trades (a paper-based system) takes place independently of price negotiation and discovery (automated on the computers of the stock exchanges), prices can be manipulated and published on the exchange, without actual trades ever taking place.

Unfortunately, because of poor liquidity and policing, broker abuses are common in Russia. Insider trading is rampant. Laws exist, but are seldom enforced. And other widespread practices have proved even more damaging. In one example, written up in Russia Review, Johnny Manglani, an Indian tailor and amateur investor claimed his Russian broker cheated him out of more than $100,000 on the Russian market. According to Manglani, Charles Shearer of the Moscow investment house Rinaco Plus called one night with bad news: Sberbank preferred stock, which Manglani had bought the previous year at $4.80 per share, was down to $1.70 and would shortly fall to $0.30. Manglani wanted to hold on, but claims that Shearer, who had not previously taken much interest in his trades, kept calling until he agreed to sell. Soon after, Sberbank preferred jumped to $2.15. This is an example of an investor not getting the best price. Other common abuses include:

Price manipulation. Because the Moscow Stock Exchange does not handle settlement, 80% of trade in Russian shares happens outside the Exchange, and investors are often left at the mercy of their broker to tell them how much a given stock really costs on a given day.

Favoring the house. Many of Russia's biggest houses do a lot of trading on their own account, known as the "proprietary book." When the traders want to get out of something, they tell their clients that it is a great buy.

Front-running. In a market as illiquid as Russia's, one big traCan send a stock's price soaring. A broker can first buy a little for the proprietary book, or for himself, then profit when the price rises on the client's purchase order. Brokers can also front-run on research reports.

Research flogging. In an up market, brokerages come out with glowing reports on companies in which they already have stakes. For example, the brokerage house Aton issued a report in September 1997 recommending the previously obscure Kazan Helicopter Factory. Aton analyst

Nadeshda Golubeva noted that Aton took a "big stake" in the factory before even beginning research. Soon after Aton's report came out, the stock plunged, and by March 1998 was removed from the Russian Trading System.

Drawing the line between ethical and unethical behavior can be difficult in an emerging market such as Russia's. In a mature market, when a bank's research department issues a buy recommendation on a formerly obscure company, clients can immediately purchase the shares through their brokers. In Russia, acquiring shares for such second-tier companies usually requires that someone physically travel to the company and buy the shares in cash from workers and other small shareholders, a process which can take weeks. Procurement can require brokers to drive long distances over dilapidated country roads, carrying millions of rubles along with a bodyguard or two. Moscow brokers need to start buying shares in advance so they will be available when the research report is published. In the U.S. this would be called front-running, but in Moscow, many fund managers are willing to pay a premium for this service.

On August 17th 1998, the Russian government defaulted on its own debt, ordered private borrowers to default on foreign loans, and abandoned its support for the ruble. The stock market, never liquid in the best of times, literally ground to a halt. From 1994 to 1998 the Russian government raised around $45 billion from selling securities, while the Russian firms issuing these securities ran up debts of another $20 billion. During the same period, an estimated $60 billion left Russia as "flight capital," finding new homes in offshore accounts in the Grand Caymans and Bermuda. (Economist, December 19th, 1998)

Many of the weaknesses in Moscow's Central Stock Exchange were symptoms of deeper faults that exposed Russia's worst inclinations. In the aftermath of the Russian privatization debacle, governments throughout Europe and Asia reconsidered the wisdom of open capital markets. Yet in Russia, it was not the concept of open capital markets that was to blame (it was their implementation through incomplete and flawed electronic trading systems. Three specific failings destroyed Russia's financial credibility:

1. *De facto* limitation of participation to a small circle of traders with inside information
2. Separation of price discovery and settlement operations, which allowed brokers wide latitude in manipulating posted market prices
3. Failure of listed firms to provide transparent information about the quality and profitability of their operations.

Had Moscow's Central Stock Exchange designed these problems out of their system prior to operation, the denouement might have been much different.

CASE STUDY IN THE AUTOMATION OF SANTIAGO'S COMMERCIAL STOCK EXCHANGE

Chilean privatisation and economic reforms were initiated under the régime of Generallissimo Augusto Pinochet. Over the 1980s these reforms steadily improved the state-controlled economy inherited from Salvador Allende'a Marxist régime. By the time that Pinochet resigned in 1990, Chile had the strongest economy in Latin America.

Shares of Chilean firms were traditionally traded in the *Bolsa Comercio de Santiago* (Santiago's Commercial Stock Exchange), which was Chile's largest and oldest stock exchange. Trading volume exploded in the 1980s with the initiation of a national pension plan that allowed workers to choose their own investments. During that period, Chile's economy grew in excess of 10% annually. An urgently needed update to the automated systems of the exchange took place from 1992 to 1994.

The exchange automated in the face of stiff competition from rival Bolsa Electrónica, a completely electronic exchange started in 1989. The Bolsa Electrónica used a computer trading system very similar to the new electronic system developed for the Bolsa Comercio, but without a trading floor, nor the accompanying expenses of a floor. In just three years it had captured 30% of transactions in securities of listed companies, where virtually all of the companies listed on the Bolsa Electrónica were also listed on the Bolsa Comercio. The Bolsa Electrónica was cannibalizing transaction volume directly from the Bolsa Comercio!

Success fomented new demands by brokers. Many brokers resided in the Exchange building, which possessed Ethernet network links providing much faster communications than existing 9600 Baud lines. They were willing to pay the Bolsa considerably more for access to Ethernet links. But linking some brokers to fast communications lines, while leaving other brokers to trade on slower 9600 Baud lines could create a caste system which would rob otherwise qualified brokers of business simply because of their location.

Even more divisive was the demand by brokers for the ability to "program trade" — to trade automatically relying on decisions made by software programs, using information gleaned from the Bolsa's electronic information feeds. Both problems presented significant challenges to the Bolsa's fairness policy, by allowing brokers who invested in information technology to place orders faster through precedence in information processing speed.

Fairness, in the sense that no trader could systematically gain an advantage over another trader by exploiting idiosyncrasies of the exchange mechanism, was particularly important in Chilean trading. Many of the traders traded on their own account, with family, rather than institutional capital. Family traders would be less inclined to invest in information technology than institutions and brokers, and without the Bolsa's commitment to fairness, would be at a disadvantage. The Bolsa was legitimately concerned that this, in turn, might drive away their business.

So important was fairness to the Bolsa that they insisted on owning the PC terminals used by brokers, and tightly controlling the software residing on those PCs, assuring that no program trading was spirited onto the PC platforms. They recently sold their new computer trading system to the Cali exchange in Columbia, which supported trading in both Cali and Bogota. In order to assure fairness, trading in Cali was purposely delayed by a fraction of a second to equalize order posting delay originating from telephone switching delays from Bogota 200 miles away.

In 1992 the Bolsa Comercio supported the following three market mechanisms for trading

1. *Open outcry:* Open outcry is the traditional method for market trading in a "pit." This market is opened from 9:00AM to 11:45 AM Monday through Friday. Here, for example, a trader enters the "stock" pit, shouts out a particular price to buy a number of shares, another trader shouts back acceptance of the price, constituting the consummation of a contract.
2. *Bulletin board:* This auction system is used primarily for thinly traded stocks, where it may take several days for bids and asks to match. This was traditionally a chalk board on which any listed company could post securities.
3. *Direct order:* This exchange allows person A to request to buy or sell directly from or to person B, but broadcasts the offer to see if there are others who would like to try to outbid either party.

The stock exchange was concerned about maintaining transparency of market activities. To this end, it supported a series of information dissemination activities to distribute information on a timely and accurate basis. The Bolsa published a series of informative brochures on various aspects of investing in the market, and on industries and firms represented in the market, including their Revista Tendencias Bursátiles (Market Trends Magazine). They issued a daily Boletín Bursátil, listing opening, closing prices, transaction volumes and so forth, along with trimester and annual reports. This information was the basis for postings in daily newspapers.

As an integral part of their system to assure market transparency, the Bolsa provided a system of computers, microcomputer terminals (in brokers and Bolsa administrative offices) and television monitors which provide up to the second information for investment decisions, drawing on the Bolsa's central databases. This terminal network provided information on price movements, price indices, income, historical and trend information on individual stocks, financial records, international price indices, interest and exchange rates, futures contracts, news and other important information.

Information from the Bolsa's central databases is communicated throughout the Bolsa, and to banks, brokers and governmental officials through a secure PBX (private branch exchange, a proprietary telephone system). In addition, the Fono Bolsa and Data Bolsa systems make price records from the Bolsa available through commercial telephone lines (through dial-up to a 700 number, which is the equivalent of a U.S. 900 number). Through this service, the Chilean telephone exchanges allowed the general populace access to the Bolsa's videotex services on televisions at home or the workplace; provided an automated voice query service for specific securities; and provided a "what if" capability through either voice or videotex to compute the value of an investor's specific portfolio. They also provided their Centro de Informatión Bursátil to handle voice queries on questions which investors could not answer through the other services.

Until 1991, electronic trading at the Bolsa Comercio de Santiago was handled on a Wang computer, which was responsible for reporting from, and, through intermediaries, posting transactions to the open outcry system. The new system was operational in January 1994, after 18 months of programming and testing by a team of four programmer/ analysts under the direction of Dr. Carlos Lauterbach. Software was tested for six months prior to going on-line, by running simulations involving two to three days of actual transactions. These simulations compressed or expanded time to assure that the systems processing capacity was adequate to reliably handle worst case scenarios. The Bolsa Electrónica and the Bolsa Comercio provided very similar capabilities to traders.

To maintain its image of fairness, the Bolsa Comercio rents out hardware and software services on the electronic exchange, forcing all transactions to be input through keyboards, and to be output through computer screens. The exchange mechanism is systematically leveled at the bottom — all clients have the same equipment and software, as well as the same communications line delays. This disallows "program trading" — i.e., the following of stock prices through electronic feeds, with posting of transactions by computer programs, expert systems, or artificial intelligence software which automatically looks for arbitrage or investment opportunities. Communications line delays are also tightly monitored, so that terminals, e.g. in the same building as the Bolsa, are not able to post their transactions more quickly than those further removed.

The electronic system captures trading that previously would have been handled through open outcry. By default, the open outcry system is fair — it is essentially self-regulating in this regard (though perhaps not efficient). It is the goal of the electronic exchange to capture that same image of "fairness" through sophisticated sets of trading policy implementations. The goal is to make the exchange systematically fair to all traders; and to avoid any features that would make the exchange systematically unfair — i.e., reward some classes of traders at the expense of other classes of traders. For example, in the U.S. issues of fairness have arisen concerning the National Association of Securities Dealers dealer-quote driven NASDAQ system, which has traditionally exhibited wider bid-ask spreads than the competing order-driven New York Stock Exchange.

The Bolsa maintains a network for broadcast of market transactions and indicators through several services:

1. continuous updates to Reuters news services through dedicated communications lines owned by Reuters
2. dial-up modem services to personal computers
3. videotex services to televisions at home or the workplace, provided by the Chilean telephone exchanges
4. an automated voice query service for specific securities
5. a "what if" capability through either voice or videotex to compute the value of a given portfolio.

In both open outcry and electronic trading, the market provides a nexus for all offers and queries brought to the market. This presents the traders with a potential bottleneck. The electronic system relies on extremely fast on-line transaction processing (OLTP) by a Tandem Cyclone computer. In data servers (i.e., computers dedicated to "serving up" data records for update or reporting) such as the Tandem Cyclone, typically several CPUs (central processing units = the computing part of the computer, of which a desktop PC computer has only one) are required to achieve the desired processing performance. The Bolsa Comercio de Santiago set as a benchmark that 95% of matchable transactions (i.e. where the bid and ask prices cross) could be completed on the Tandem within one second. In order to do this, the Tandem computer needed to be run at 40% of its peak load capabilities (a statistic established by the manufacture Tandem Computers, Inc.). This required a

Tandem computer with six CPUs, to support 30,000 transactions per day (an average of 10 transactions per second) from 600 dedicated terminals in the offices of agents and brokers. In addition, the system needed to support the broadcast of stockmarket information (essentially a full screen electronic ticker tape) through dial-up modem connections, and voice announcing provided through the equivalent of U.S. 900 numbers available through the telephone companies. Tandem processing was estimated to be linearly scalable to about 300% of this performance, at a maximum configuration of 16 CPUs.

In traditional markets (including the open outcry market at the Bolsa Comercio) the matching bottleneck is handled by breaking out trading by specific assets — i.e., certain stocks are traded only in one pit. This was tried in the electronic exchange. It failed to yield improvements in matching speed due to time clustering of trades in any given security or industry during the day, usually around the time of release of critical information such as financial reports. 1992 transaction growth of 20 to 30% annually threatened to outstrip the Tandem's processing capacity in around five years. Yet they knew that the New York Stock Exchange, using Tandem computers, was able to maintain similar performance standards, applying a more complex trading policy, for transaction volumes in the 1 million to 10 million transactions per day—30 to 350 times the current volume on the Bolsa Comercio.

One of the most critical decisions in systems development for software spanning several machines is the scheme for splitting the workload between computers. The Bolsa had chosen to dedicate matching activities to the Tandem, broadcast activities (to Reuters, the Fono Bolsa, and other news reporting services) to the Sun, and transaction acquisition and real-time reporting to the PCs located in brokers' offices. Several factors are important in the Bolsa Comercio system:

1. The constrained machine resource is the Tandem Computer, which is the nexus for transaction matching. The Tandem should not be responsible for broadcasting information on trades, volumes, and so forth. Broadcasts are required by the dial-up services offered over the telephone lines, as well as continuous updates to Reuters, and other news services.
2. The constrained resource for systems development are the programmers, who are skilled in COBOL programming, but know little of event triggering (e.g., posting an offer), or of software development for Intel PC client terminals.
3. The performance and reliability of the Intel PC client terminals may be uncertain, since they are in users' hands. Thus their responsibilities need to be limited to reporting status of offers and trades, and to input of transactions (offers or queries).

4. Cost issues were important. The six CPU Tandem cost US$1 million, depreciable over four years (mainly technological obsolescence). The client terminals were i286 or i386 PCs costing around $1,000 per terminal.

These constraints suggest that broadcast of information — such as required by Reuters or the telephone voice and data broadcasts of market information — can be handled by neither the Tandem nor the terminals. For this reason, an intermediate tier of Sun workstations ($5,000 per terminal) were installed to handle two tasks: (1) format updates to the market for broadcast to the telephone companies and to news services such as Reuters, and (2) reliably format transactions coming from the client terminals for forwarding to the Tandem. The Sun workstations were wired to both the Tandem and to the client terminals in pairs to assure that if one Sun went down, another "hot standby" was ready to complete the transaction in less than one second (i.e., in "real time"). Each Sun workstation could handle transactions from around 100 terminals.

All of this technology was employed in the achievement of three computational performance targets:

1. 95% of matching transactions (on Tandem) completed in less than 1 second
2. Sun batching and reporting introduced 200 milliseconds (.2 second) delay
3. No PC software module greater than 10Kbytes in size (which translates into less than 3 seconds download to update a PC terminal's software, done at time of logon)

The Tandem is mainly used as a data server, though some of the server programs (coded in COBOL) provide information search and formatting services. Since reliability, accuracy, security and integrity are paramount to this function, the Tandem implements software and data mirroring. Software and data mirroring are services that Tandem is famous for, and one of the reasons that they are the primary supplier of computers for stock exchanges and other mission critical applications where any downtime is unacceptable.

The majority of software and processing in the client terminals supports the graphical user interface, running under Windows. This is programmed in the ECU language, which is compiled (i.e., translated) into a very compact intermediate language (which can be quickly downloaded across 9600 baud communications lines if needed), and is interpreted (i.e., run) by the kernel.

The Sun workstations contain no custom code. They use packaged software for information broadcasting, for managing the client terminals, for performing transactions for posting to the electronic exchange (on the Tandem), and for managing the multiple sessions running simultaneously. By not using custom code on the Sun workstations, and by performing only rudimentary services, well suited to a Tandem server on the Tandem, the job of updates, maintenance, extensions and revisions to the system are kept mainly on the client terminals. And the client code is loaded to the Tandem, where it is automatically downloaded to each client whenever it logs on.

The Sun workstations:

1. display amounts on the bulletin board in the sala de ruedas used for displaying transactions
2. handle routing of transactions; storing and forwarding of transactions for integrity
3. randomize the order of transactions in the queue to be forwarded to the Tandem for matching, and randomize the order of transactions in the similar queue to be broadcast to traders. This function is provided by the stock exchange to insure fairness in trading, and to thwart attempts at front-running.

The Sun workstations add a 200 millisecond per transaction delay (i.e., 20% of total one second target time for processing a transaction) due to storing and forwarding, but this is considered worthwhile, since during peak loading, the Tandem is freed for search and matching functions.

TEN LESSONS THAT INTERNET AUCTION MARKETS CAN LEARN

In all three cases of stock market automation presented in this paper, the move towards automation, and linkage into a wide-area order gathering and information dissemination network improved liquidity, extending the reach and participation in the market. Each of the exchanges used information technology to rewrite its trading geography, making off-floor transactions more appealing than on-floor transactions. Yet none of the moves off-floor was without costs and problems. These provide valuable insights for e-Commerce's automation of retailing. This section summarizes ten lessons from securities exchange automation, with examples from Santiago, Moscow and Shanghai stock markets.

ELECTRONIC AUCTION MARKET ADVANTAGES (LESSONS 1 THROUGH 5)

Lesson 1: *Customers are attracted to electronic auction markets because they provide greater liquidity than traditional markets; ceteris paribus this greater liquidity results directly from greater geographical reach provided to commercial transactions by electronic networks.*

A liquid market is one in which traders can quickly satisfy their demand to buy or sell specific commodities. Traders (i.e., the customers of securities markets) fall into two broad groups: those who participate in order to buy the product (usually called value motivated traders; and those who hope to resell the product in the same market, taking advantage of supply-demand imbalances (usually called speculators. Speculators can be valuable liquidity providers, but can also undermine efficiency and orderly trading by distorting the reflection of underlying value of a product in the market.

The importance of a liquid market was not lost on Shanghai Stock Exchange officials when they automated their exchange. In their case, raising capital was of secondary importance, since the Chinese save 40% of their earnings, which was the basis for much of the lending to State enterprises. Rather they wanted their exchange to attract foreign capital to these State enterprises, while assuring that any savings transferred to investments from the Bank of China (the major lender to the State sector) were still invested in the State sector. They accomplished this by listing State Owned Enterprises (SOE) on the exchange, and insuring that trading was easy, attractive and accessible to all Chinese people. This insured liquidity, by attracting a large base of active traders in listed securities by applications of computer and communications technologies appropriate to China's populace. In contrast, the Moscow Central Stock Exchange kept trading in the brokerage houses as a result of their inability to provide network connections to the people. Further unsatisfactory outcomes may be attributed to this failure to provide widespread "geographical reach." The problem was even more pronounced in the Russian voucher market, which offered no automation whatsoever.

Liquidity becomes important in product markets with large trading volumes (e.g., for raw materials and intermediate production factors). These products happen to currently provide the most lucrative electronic commerce markets. The liquidity required in any particular market depends heavily on the lead time required for its use. In financial markets, this can be extremely short. In tangible goods, it may be much longer (e.g., concrete and steel contracted for a highway which will be constructed over the next decade need not be traded more than once every few months for buyers to consider the market sufficiently liquid).

It is important to note that the traders who can potentially offer the most liquidity to a given market are not necessarily those traders who are either able to or are willing to be continuously present in a given market. This last premise is particularly important in assessing the impact of automation on markets, since advances in computer and telecommunications technology have obviated the need for physical presence at a trading location. This latter aspect of automation, along with the possibility of maintaining a quasi-continuous presence using the vehicles of automated (intelligent) agent software to continuously monitor and transact in an automated market, is perhaps the greatest source of impact of automated markets.

Lesson 2: *Electronic auction markets can more efficiently discover the best price at which to trade in a product.*

Buyers and sellers prefer — and will choose to trade in —a market that insures that the best price is discovered which can maximize buyers' and sellers' preferences. One reason to trade in an auction market, rather than simply search for a product in a retail market, is the potential to discard the "price-taking" assumption of the retail market. This forces the seller to give up the producer surplus—the amount received in excess of a particular buyer's valuation of the good—to the advantage of the buyer. Such a market will attract more sellers, because it can ensure them a better deal. It will attract more buyers, because of the greater liquidity arising from there being more buyers.

Best price discovery should be improved with increasing liquidity. The greater the liquidity of trading, the lower will be the producer and consumer surplus, because trading can potentially occur at more points along the supply and demand curves. All three cases show how modern computer and communications technologies can expand the number of customers (traders) participation in the market, by greatly reducing the need for customers to be at a specific location at a specific time to trade. Where such limitations on auction location and time were imposed (as in the case of Russian industry vouchers), customers were greatly disadvantaged.

Lesson 3: *Electronic auction markets can, at low cost, provide exceptional levels of transparency of both market operations and of products quality.*

Transparency of a market implies that there exists some public source of information about products traded, and the activities of the market itself. This information must be current, address product quality, and must be credible. In financial markets, audited financial statements and continuous news reporting provide a reliable source of information.

Transparency of trading is another way of describing the demand functions of participants in the market. Three broad levels of transparency of trading on any single security are possible (with finer distinctions within each one of these levels):

1. Only the quotes of market makers may be broadcast to the trading public at large. This corresponds to the closing prices posted in newspapers,

2. The price and volume on the last trade may be posted. This corresponds to the ticker information broadcast throughout the day, or
3. The price and volume on each order can be posted. This corresponds to a market maker's or broker's order book.

Of the three markets described in this article, only the Santiago Exchange provides transparency of both fundamental business information in the firms whose securities are traded, and of market trading. All three exchanges provide trading transparency at level (1); the Santiago and Shanghai exchanges provide transparency at levels (2) as well. It is rare for brokers or market makers to reveal their order books, though. Moscow's stock market, because it processed only about 20% of trades in listed securities, lost the confidence of investors in part because market prices did not adequately reflect supply and demand.

Lesson 4: *Electronic auction markets are more efficient than traditional markets. This efficiency allows them to better provide information required to correctly price assets traded in the market.*

Efficiently is related to transparency in the sense that it has to do with the dissemination of information. Efficient market prices will accurately reflect underlying value, supply, and demand for a product. In efficient markets, the price will be quickly influenced by new information.

The Moscow Exchange again provides an example of problems arising from failure to rapidly incorporate new information into prices. In the Moscow Exchange's case, this arises because the majority of trades (around 80%) take place off the floor, at transaction prices which are not posted. As a result, posted market prices tend to be ignored by traders. They also tend to be lower than the estimated book value of the shares. This illiquidity is reflected in bid /ask spreads that are often an order of magnitude apart, whereas in most exchanges, it is a fraction of the smallest currency unit.

Lesson 5: *Electronic auctions can provide a market that, ceteris paribus, offers services at a lower transaction cost.*

Markets provide a specific bundle of services. Where consumers have a choice of two otherwise equivalent bundles, they will do business where transaction costs are lowest. Traditional securities markets did not need to be significantly concerned about transaction costs because an exchange's dominant position in trading a listed security, the tendency of liquidity to draw in more transactions, and regional loyalties combined to limit the choice of venue offered traders. With the drop in price of hardware and software for creation of an exchange, more and more markets can enter the competition for order flow in the same securities, competing on convenience and transaction cost. Thus Bernard Madoff began an electronic exchange in securities over a decade ago that now does over 10% of the volume of the New York Stock Exchange on listed securities. More recently, E*Trade has opened its own exchange, competing for similar stocks, but offering its own on-line order capture.

The situation highlighted in the Bolsa Comercio de Santiago case involves the challenge to its dominance over Chilean market trades by the Bolsa Electrónica, a completely electronic exchange started three years earlier. The Bolsa Electrónica used a computer trading system very similar to the electronic system developed for the Bolsa Comercio, and in just three years it had captured 30% of transactions in securities of listed companies (that figure had risen to nearly 60% by the late 1990s). The Bolsa Electrónica was cannibalizing transaction volume directly from the Bolsa Comercio, and was able to process transactions less expensively because it was able to do away with the trading floor, and its associated maintenance and rental expense, displays, labor, and so forth.

Not being able to compete on cost, the Bosa Comercio chose to improve service in order to attract business. The case focuses on the trade-offs between policy (e.g., fairness and transparency) of the market, and the technology offerings required to make the market attractive (e.g., Ethernet communications lines and program trading). Its operations emphasized fault-tolerance and on-line transaction processing and effective information dissemination systems to provide real-time transaction and summary data.

ELECTRONIC AUCTION MARKET CAVEATS (LESSONS 6 THROUGH 10)

Lesson 6: *Customers will abandon a market that is not perceived as fair, even though they may initially profit from "unfair" transactions in that market. By distancing customers from the traders in a market, they can provide a false sense of legitimacy to a market that allows unfair and opaque trading practices.*

Trading is fair if one trader cannot systematically profit from another trader. Fairness is related to information efficiency of a market in that it requires that information about the security, as well as information about offers, sales and the trading environment, be available to all traders in the market. Even so, fairness may be difficult to define in general. Most traders assume that they can outperform the market, and thus assume that they can systematically profit from less informed (and presumably less clever) traders.

Markets need to insure fairness in order to attract trading (otherwise the disadvantaged traders would not trade in the market, leaving the remaining traders without traders from which to systematically profit). The resulting loss of liquidity is likely to drive out even traders who have initially benefited from "unfair" transactions.

Grossman and Stiglitz (1982) showed that in fact market's could not be efficient and fair in the sense that all traders are equally informed (otherwise trading would stop). What is needed is the perception of fairness, in that any differences in success in market trading are presumed to be due to trader risk-preferences, luck, and competing philosophies about what drives performance in a particular industry or stock.

Markets need to adapt them to the business environment around them, and fairness must be interpreted in a local context. Russia's markets provide an example of the difficulties in defining fairness in any universal sense. Broker abuses, insider trading and market failures are common in Russia. Some abuses (e.g., broadcasting prices via the exchange system, but completing the transactions at a different price in a private transaction) could be avoided by greater integration of the brokerage community into the market making system. But some problems arise from the inefficiency and illiquidity of the market. Brokers may receive little information on firm assets or operations in Russia, and thus are at greater risk of misrepresenting the investment potential of a product. Illiquid markets make it difficult to assure that the next transaction takes place at a price close to the prior transaction, or to know whether the difference is due to manipulation. The Russian markets provide a telling example that your markets must monitor, enforce and insure fairness in trading

Lesson 7: *Electronic auction systems must manage all aspects of trading activity, from initiation to settlement and delivery. Markets that fail to integrate both price discovery and order completion (settlement) into their operations can encourage unfair trading behavior, and opaque trading practices.*

This lesson is drawn from abuses at the Moscow Exchange where credibility of prices suffers significantly from failure to accurately record traded prices. In the Moscow Exchange's case, this arises because the majority of trades (around 80%) take place off the floor, at transaction prices which are not posted. As a result, posted market prices tend to be ignored by traders. They also tend to be lower than the estimated book value of the shares. Bid/ask spreads (the difference between the price at which a share can be sold on the exchange, and that at which it can be purchased) may be an order of magnitude, whereas in most exchanges, it is a fraction of the smallest currency unit.

Market prices may be set through two classes of mechanisms — (1) dealer quotes, and (2) price matching of bid and ask offers. The former are called quote driven, and the latter order driven markets. In an order driven market, the stated price that a market sells or buys a particular commodity is typically the price at which the last trade took place. In a quote driven market, quotations simply indicate that a particular market maker is willing to trade at specific bid or ask prices. Quote driven markets are similar to retail stores in that traders coming to the market must take the prices quoted. Retail stores typically serve only buyers, whereas quote driven markets serve both buyers and sellers.

Lesson 8: *Because the delay in price response may be significantly faster completion and posting times, there is greater potential for feedback loops and instabilities that are a threat to orderly trading, and to fair and efficient pricing of assets traded in the market.*

Traders derive considerable information from price movements in markets. Thus prices should not be overly influenced by imbalances in supply and demand (prevention of shortages and price fluctuations is one of the main reasons for using a market to trade). The last price at which a security was traded is close to the price at which they can trade. Where trading is not continuous, e.g., when a stock exchange closes at the end of the day, orderly trading may be insured by artificially setting the opening price of the stock to clear the outstanding buy and sell orders. All three of the exchanges in this article set opening prices to clear outstanding orders. In the case of the Shanghai exchange, this may often be considerably different from the last trade price of the prior day.

Market operations must assure that prices do not vary greatly unless this truly reflects an ongoing trend or change in fundamental value. Specialized market-making functions and trading rules such as circuit breakers (which stop trading if an excessive drop in prices is detected) promote orderly trading on the underlying value of securities.

Lesson 9: *Electronic auctions may foment unfair trading practices through different relative speed of service through different parts of its network linking trading to customers.*

The issue of speed is a subtle one, but is intrinsically tied to fairness and transparency. It appears in three places in trading: (1) order placement, (2) information dissemination, and (3) matching. Absolute speed will change continually, as new technologies are introduced. It is important that the speed offered by a market be competitive with competing markets on these three dimensions.

An example of the importance of speed in order placement is provided in software of the Commercial Stock Exchange of Santiago when it was installed in Cali and Bogata, Columbia. To insure fairness of trading, the Bogata communications line actually had to be slowed down so that Cali traders couldn't trade in advance of Bogata traders on the same information. A similar situation arose inside and outside the Exchange building in Santiago, where the local network access was considerably faster than city-wide access.

Santiago's market was sensitive to speed of information dissemination as well. The market operators recognized this, and provided several innovative ways of accessing real-time quote information, including a telephone dial-up system. At the opposite end of the spectrum was the Moscow exchange, where many stocks traded once or twice a week. In this case, daily newspapers were more than adequate for reporting.

Matching proved to be the major bottleneck in the Bolsa transaction processing. With the increase in velocity and sophistication of trading — particularly the technology for program trading — comes increasing sensitivity to matching delays. Program trading, and exceptional variability of volumes could significantly degrade matching speed at peak times, thus degrading market performance on other parameters such as best price discover and orderly trading.

Lesson 10: *Order driven electronic auction markets demand that the market clearly define when a sale has been made.*

In an order-driven market, customers place buy and sell orders in the market, in hopes of their being matched with other customers' orders at the best price. At the point the orders enter the market, the traders lose control over the price negotiation process. Naturally, they will be concerned with the system that the market uses to complete their transactions. The algorithms used by the marketplace need to be clearly defined to the market's customers in order to attract trading.

In order-driven securities markets, order management can become very complex. In a survey of 50 electronic trading systems in 16 countries, Domowitz [1992] found: 11 trade priority rules (e.g., best price has highest priority, time priority is first-in first-out) ; 7 levels of price discovery rules for information search and matching (e.g., manual exposure of orders to a market maker for price improvement; anonymous, direct negotiation); and 12 classes of information which define the transparency, type and amount of information which is disseminated to traders (e.g., high and low price, best bid and offer, and quantities available at these prices). Markets can give customers some choice over the determination of the time of sale (and thus sale price, quantity and so forth) by allowing a variety of orders (limit, stop limit, market, margin, and so forth.

NOTES

[1] *Asiaweek* (1997) Feb. 28th 1997, p. 62

[2] *Economist* (1997) June 14th 1997, p.124

[3] *Economist* (1998) Capitals of Capital, Survey, May 9th, 1998

[4] Delays are not just introduced by the length of wire (or fiber), which given the speed of light would be negligible. Rather, switching delays and error correction (which may require resending information, and transmitting signals in two directions for confirmation) introduce the most significant delays. Indeed, the modified version of Bolsa Comercio's electronic trading system that was sold to the Cali exchange in Columbia had to support a large number of transactions from both Cali and from Bogota 200 miles to the northeast. To accommodate this, signals at Cali were artificially delayed to allow the Bogota signals to "catch up," insuring fairness in trading, and diminishing the possibility of "front-running."

[5] Linearly scalable means, roughly, that if the Bolsa buys three Tandem CPUs, it will have a machine that processes roughly three times as many transactions in the same time interval.

[6] Mirroring means that copies of the same data or program code are maintained in two places, managed by the same system. This is necessary for mission critical systems, where information services cannot afford to be lost, as in the case of market transactions. It is a critical feature in providing "fault-tolerant" systems—i.e., systems which can recover quickly from faults, whether they be system or user induced.

REFERENCES

Asiaweek (1997). Feb. 28th 1997, p. 62

Domowitz (1992). *Liquidity and Trading Rules*, Northwestern University Working Paper

Economist (1997). June 14th 1997, p.124

Economist (1998). Capitals of Capital, Survey, May 9th, 1998

Economist (1998). The cash don't work, December 19, 120-112

EDI Forum: Voucher Privatisation Funds in the Russian Federation," http://www.tomco.net/~edinp/fsu/funds.html;

Euromoney (1994) Bulls go wild in a China shop, October, 56

Euromoney (1996). Varying fortunes of China chips, March, 15.

Euromoney, (1996). Varying fortunes of China chips, March.

Federal Commission for the Securities Market (1997). "Report at the All Russian Conference of Professional Capital Markets Participants," October 1996; and *Moscow Times*, "FSC to Create Regional Self-Regulatory Network," July 26, 1997, 14.

Grossman, S and J. Stiglitz (1982). On the Impossibility of Informationally Efficient Markets, *American Economic Review*, 72(4).

Harris, L.E. Liquidity (1991). Trading Rules, and Electronic Trading Systems, *Monograph Series in Finance and Economics*, New York University Salomon Center.

Kommersant, No. 10, March 22, 1994, "Better to Buy Gazprom", 32-34.

Kommersant (1994). "Duma Debates on Privatization", No. 15, April 26, 37-38.

Kommersant (1994). "Voucher Privatization is Not Likely to Impact the Gas Industry", No. 3(1), 33-35.

Kommersant, (1994), "Privatization of Voucher Investment Funds in Reflection: Most Important is Not Victory, but Participation", No. 5, February 15, 43-51.

Moscow Times(1997). "VegaTech Seeks Market Serving Market-Makers, June 17, III.

Moscow Times, "VegaTech Seeks Market Serving Market-Makers, June 17, 1997, p. III.

Potter, P. B. (1992). Securities Markets Opening to Foreign Participation, *East Asian Executive Reports*, April, 7-9.

Rinaco Plus Brokerage House Web Site (1999). "Equity Index Methodology." http://feast.fe.msk.ru/infomarket/rinacoplus/

Roman Frydman, Katharina Pistor and Andrezej Rapaczynski (1995). "Investing in Insider-Dominated Firms: A Study of Voucher Privatization Funds in Russia," *Oesterreichische Nationalbank Working Paper 21*.

Shama, A (1997) Notes from Underground as Russia's Economy Booms, *Asian Wall Street Journal*, 30 December 1997. National income statistics are reported in *Asiaweek*, 16 January 1998, p. 51

Thomas, W.A. (1993). Emerging Securities Markets: The Case of China, *Journal of Asian Business*, v. 9(4) , 90-109

Whitehouse, M. (1998) Shortchanged on the Stock Exchange, *Russia Review*, May 8, 1998, *The Moscow Times*

Whitehouse, M. (1998, May 8). Shortchanged on the Stock Exchange, *Russia Review*.

Yergin D. and J. Stanislav (1998). *The Commanding Heights*, p. 282, New York: Random House

Chapter 7

The Societal Impact of the World Wide Web – Key Challenges for the 21st Century

Janice M. Burn
Edith Cowan University, Australia

Karen D. Loch
Georgia State University, USA

INTRODUCTION

Many lessons from history offer strong evidence that technology can have a definite effect on the social and political aspects of human life. At times it is difficult to grasp how supposedly neutral technology might lead to social upheavals, mass migrations of people, and shifts in wealth and power. Yet a quick retrospective look at the last few centuries finds that various technologies have done just that, challenging the notion of the neutrality of technology. Some examples include the printing press, railways, and the telephone.

The effects of these technologies usually begin in our minds by changing the way we view time and space. Railways made the world seem smaller by enabling us to send goods, people, and information to many parts of the world in a fraction of the time it took before. Telephones changed the way we think about both time and distance, enabling us to stay connected without needing to be physically displaced. While new technologies create new opportunities for certain individuals

Previously Published in *Social Responsibility in the Information Age*, edited by Gurpreet Dhillon,

UNIVERSITY OF
WOLVERHAMPTON
HARRISON LEARNING CENTRE

16/12/04

11:33 am

Item:Computer ethics and professional responsibility
23223219

Due Date: 14/01/2005 23:59

Item:VB.NET for developers
22625534
Due Date: 14/01/2005 23:59

Item:Learning Visual Basic .NET
22954775
Due Date: 14/01/2005 23:59

Item:Ethical issues of information systems
22589902
Due Date: 14/01/2005 23:59

Thank You for using

Self Service Issue

Please return or renew by date quoted above. (Telephone renewals: 01902 321333)

or groups to gain wealth, there are other economic implications with a wider ranging impact, political and social. Eventually, as the technology matures, social upheavals, mass migrations and shifts in economic and political power can be observed. We find concrete examples of this dynamic phenomenon during the Reformation, the industrial revolution, and more recently, as we witness the ongoing information technology revolution.

Before the Reformation, the church controlled an effective monopoly on knowledge and education. The introduction of the printing press in Western Europe in the mid-15th century made knowledge and ideas in book form widely available to a great many more people. Printing hastened the Reformation, and the Reformation spread printing further. By the early 16th century, when Martin Luther posted his 95 theses on the castle church, the political movement was well underway. The printing press changed the way in which we collected, transmitted, and preserved information prior to that time. Mass production and dissemination of new ideas, and more rapid response from others were instrumental in launching a worldwide social phenomenon.

Dramatic changes in the economic and social structures in the 18th century characterized the industrial revolution. Technological innovations were made in transportation and communication with the development of the steam engine, steam shipping, and the telegraph. These inventions and technological innovations were integral in creating the factory system and large-scale machine production. Owners of factories were the new wealthy. The laboring population, formerly employed predominantly in agriculture, moved in mass to the factory urban centers. This led to social changes as women and children were introduced into the workforce. Factory labor separated work from the home and there was a decline of skilled crafts as work became more specialized along the assembly line.

The inventions of the telegraph and telephone dramatically changed the manner in which we conduct business and live our daily lives. They allowed the collection, validation, and dissemination of information in a timely and financially efficient manner. More recently, we are experiencing the information technology revolution, led by the introduction of computers. The rate of change has accelerated from previous times–with generations of technology passing us by in matters of months rather than decades. We are witnessing significant shifts in wealth and power before our eyes. Small start-up high technology and Internet companies, and their young owners, represent a very wealthy class–and an extremely powerful one. Small countries such as Singapore and Ireland, through the strategic use of information technology and aggressive national policy, have transformed their respective economies and positioned themselves in the competitive global economy.

The Internet, a complex network of networks, is frequently spoken of as a tool for countries to do likewise. The Internet removes the geographical and time limitations of operating in a global economy. The banking industry has been revolutionized with Internet banks who can collect, validate, and disseminate information and services to any people group–internal to the organization and external to its customers-in a timely and financially efficient manner. Similar scenarios exist in the worlds of retail, healthcare, and transportation.

There is an underlying assumption in the popular belief that the Internet may be the savior to the developing countries of the world. Such thinking is dependent on a single premise: the belief that access to information gives access to the global marketplace, which in turn leads to economic growth. Information is power; knowledge is wealth. The vehicle for access is information technology and communications infrastructure (ITC). Mohammad Nasim, the minister for post and telecommunications in Bangladesh, one of the poorest countries in the world, restated the premise, saying, "We know full well how important a role telecommunications play in a country's economic development" (Zaman, 1999). The converse is also true. Lack of IT access leads to an increased inability to compete in the global marketplace, which leads to further economic poverty. What we are witnessing is therefore either an upward or downward spiral phenomenon. This raises some interesting and important questions for society, such as: What is the current information access through the Internet? Who are the "haves" and the "have-nots" of information access? How can the Internet address the societal challenges?

This paper attempts to address these questions and related issues. In the first section we document the current state of information technology diffusion and connectivity, and related factors such as GDP, population density, and cultural attitudes. The second section examines more fully the question of who comprises the "haves" and the "have-nots" so frequently mentioned. Across and within country comparisons are made, noting in particular disadvantaged groups, urban vs. rural communities, and women and children as groups that are frequently forgotten, but who are vital to true transformation to a global information society. The third section offers some concrete suggestions as to how the Internet may be used to address the growing gap between those who have and those who don't. We report some country examples which illustrate both the progress and the magnitude of the challenge as societies, governments, and other key change agents attempt to redress the problem. Finally, we make two observations. One is that for those who don't have, there is little demand to have, as well. This is in large part explained by the second observation, which is that a multilevel complex challenge must be overcome in order to leverage technology-based services, such as offered by the Internet, as a sociological tool to reduce economic disparity. We challenge the reader to look inward for each one's individual responsibility in this big picture.

INFORMATION TECHNOLOGY ACCESS

In 1995, the world IT market as measured by the revenue of primary vendors was worth an estimated US$527.9 billion. Between 1987 and 1994, its growth rate averaged nearly twice that of GDP worldwide. It was particularly high in Asia climbing from 17.5% to 20.9% of world share during that time. Nevertheless this strong growth did little to redress the geographical imbalance in the world IT market—markets outside Asia and the OECD area (ROW) accounted for only 4% of the world total.

From a world population of 5.53 billion, ROW accounts for 82.6% of the total population yet from a world GDP of US$25,223 billion, ROW accounts for only 19.2% (decreasing >2% over the last 7 years) and from a total IT market of US$ 455 billion, ROW accounts for only 8.4%. See Figure 1.

The IT market has remained concentrated within the G7 countries at around 88%, with the United States accounting for 46% of the market. In terms of installed PC base the US was by far the world leader with 86.3 million units well ahead of Japan (19.1m), Germany (13.5m) and the UK (10.9m). In the US this averages at 32.8 PCs per 100 inhabitants. The Internet now reaches into every part of the globe with the number of host computers connected to the Internet increasing from 3.2 million in July 1994 to 6.6 million in July 1995, 12.9 million in 1996, 16.1 million by January 1997 and 29.7 million by January 1998 (Network Wizards). This is more than a tenfold increase since July 1993 as shown in Figure 2.

Figure 1: Share of OECD member countries in world population, GDP and IT market, 1987–1994

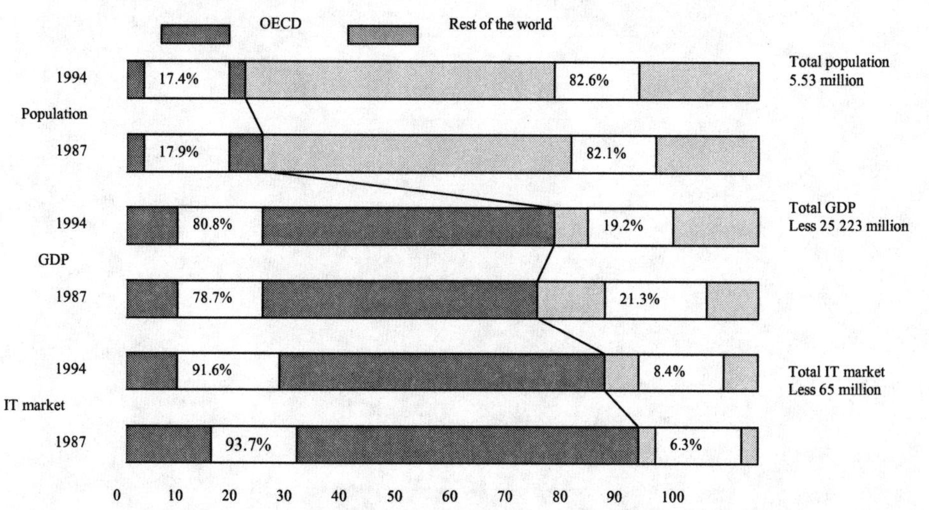

Source: *World Bank* based on IDC (1995b), *World Bank* (1995), and *OECD* (ANA Database)

Recent estimates indicate that some 90 countries, just under 5 million machines and some 100 million users worldwide are connected to the Internet (NUA Internets Survey, 1998). However, Internet hosts per 1 million inhabitants by country income show huge differences between the rich and the poor, with 31,046 hosts for the highest income countries and only 9 per million

Figure 2: Internet host computers (millions)

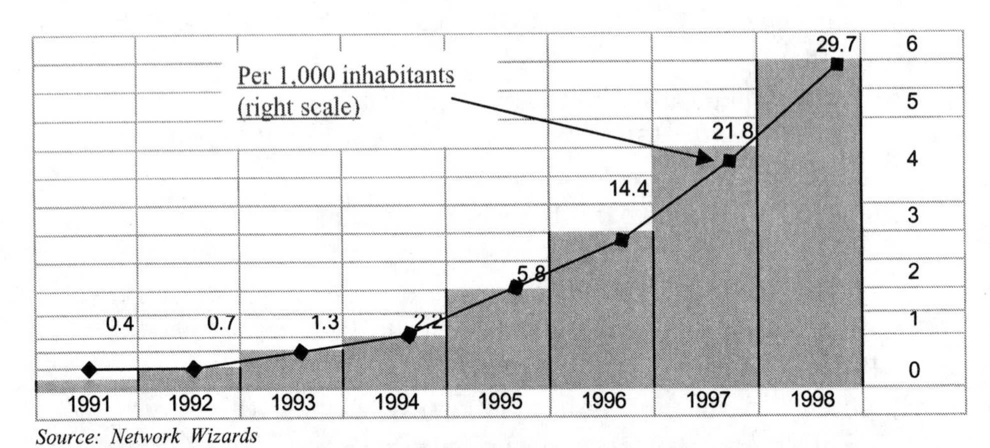

Source: Network Wizards

Figure 3: Number of PCs per 100 white collar workers

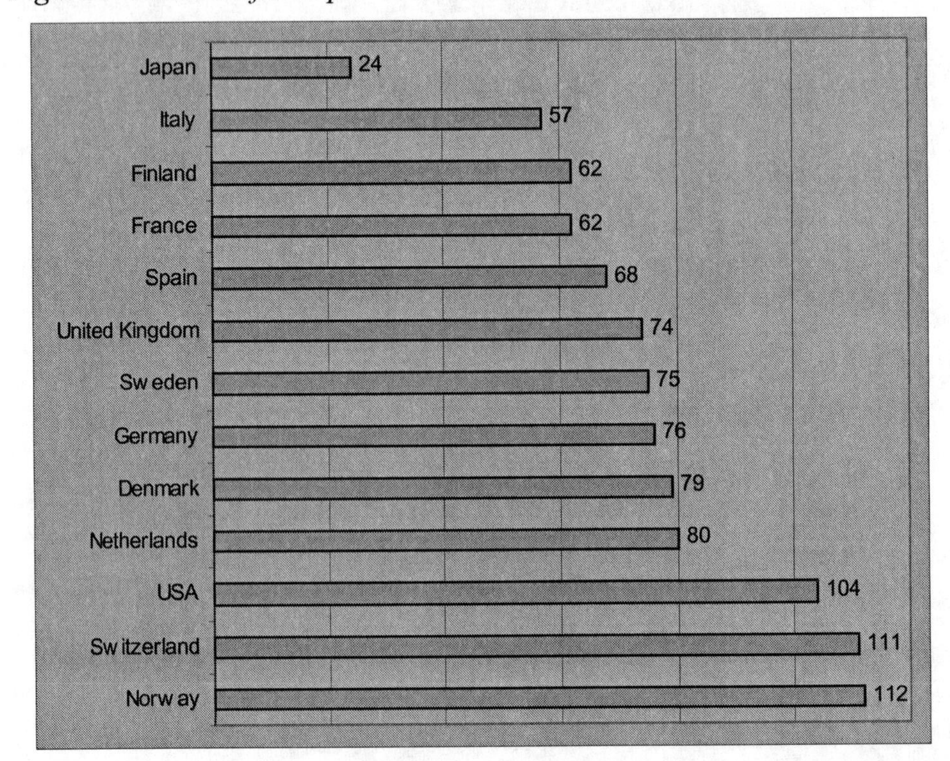

inhabitants in the poorest. The level of LAN implementation differs significantly across countries, with the US accounting for 55%, Western Europe 32% and ROW only 13% of the installed base of LAN servers.

This has to be examined at two levels: the rates of PC diffusion and connectivity. In terms of the number of corporate PCs per 100 white-collar workers, leading countries such as Norway, Switzerland and the US have more than 100, major Western European countries 60-80 and Japan only 24 (see Figure 3). As for PCs connected to LANs, 64% of corporate PCs are on a network in the US but only 21% in Japan (Dataquest, 1995). Corporate cultures in Asia may be less conducive to online management.

Access to telephone service is a good indicator of the state of a country's telecommunications infrastructure as this plays a large role in accessibility to information. More than 90% of households in high-income countries have a telephone line (and some have more than one), whereas only 2% of households in low-income countries are similarly served. Of 950 million households in the world, 65% of the total do not have a telephone. Figure 4 shows the distribution of telecommunications against wealth.

The technology gap is strikingly apparent in telephone usage, where consumers in the United States make an average of 2,170 calls per inhabitant annually, which converts into just under seven calls a day. Only Canada and Singapore come close to the American average; Canada because of the similarity of culture and technological deployment, and Singapore by virtue of the heavy concentration of business within the small city-state. The United States' use of the telephone remains approximately three times higher than the European, Japanese and Australian averages, which seem to be clustered at between the 600-800 call per inhabitant level.

Figure 4: Lines per 100 inhabitants in relation to GDP

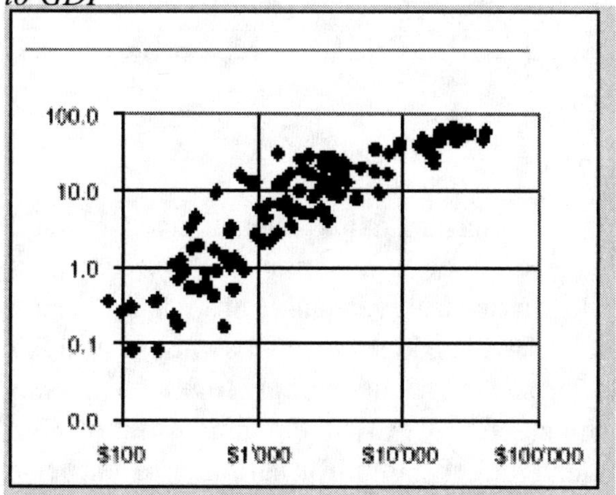

Source: International Telecommunication Union, 1998

The difference between the United States and the Latin American and some of the Asian countries is even more striking. The average American makes 10 times as many calls as the average Mexican, 20 times as many calls as the average Chinese, and 40 times as many as the average Indian (see Figure

Figure 5: Number of calls per capita by country

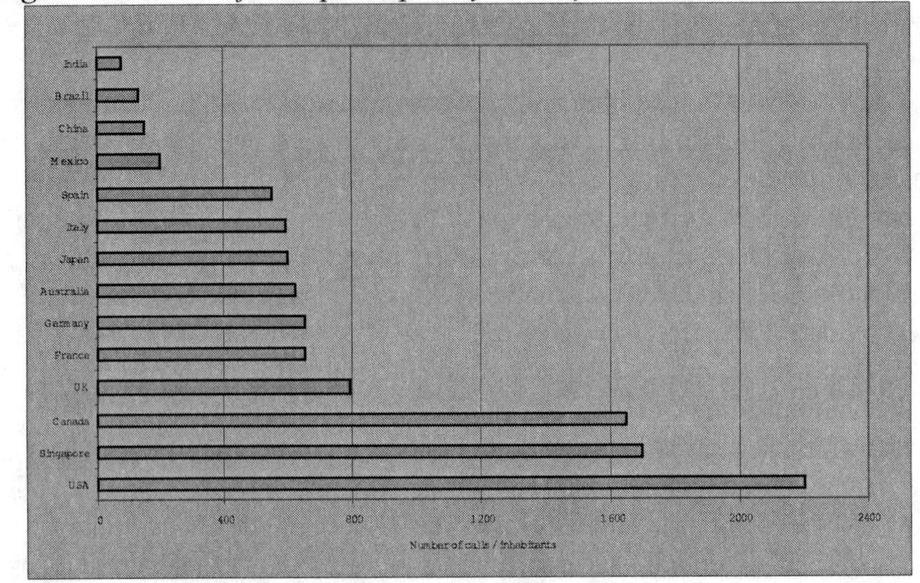

Source: International Telecommunication Union, 1998

Table 1: Worldwide indicators

	1995 US$ gdp pc	School life expectancy	Adult (F) illiteracy	Economic Rural Activity %	% access to sanitation
USA	26037	15.8	3.1	59.9	*
Japan	41718	14.8	*	50.0	*
UK	18913	16.3	*	52.8	*
Australia	20046	16.2	*	48.1	*
China	582	*	27.3	72.9	7
India	365	*	62.3	*	14
Philippines	1093	11.0	5.7	49.0	67
Argentina	8055	3.8	3.8	41.3	2
Vietnam	270	*	8.8	74.1	15

* not available

Source: UN statistics, 1997

5). As the developing countries make greater inroads into extending their networks and their inhabitants succeed in integrating the telephone more into their daily lives, it is to be expected that their telephone usage will eventually start to catch up to that of the more developed countries, but it will undoubtedly take some time to do so.

While the technology invasion has offered developing countries amazing opportunities to leapfrog over stages of growth in their programs for industrialisation and advancement, the drive for information can often occur only at the expense of other basic infrastructure needs which are regarded as norms for advanced societies. Illustrative of these trade-offs are countries who are currently making major investments into their ITC infrastructure, as shown in Table 1. China aims to enter the 21st century as an information economy yet has an average GDP which is only 1/50th of the US; Argentina has a school life expectancy of less than 4 years compared to over 16 in Australia, and India boasts a female adult illiteracy problem of 62.3%. The statistics are even more alarming when comparisons are made with rural communities, where only 7% of the rural population in China and 2% in Argentina have access to sanitation.

The Haves and the Have-Nots

The haves and have-nots are generally differentiated based on a variety of factors such as income and education levels. We generally think of the haves and the have-nots from the perspective of the international arena, dividing countries into two large categories: developed and developing, with the greater proportion of countries considered developing. There is a tight coupling between the ITC infrastructure of a country and its income status. It comes as little surprise then that despite rapid growth of the Internet, some 97% of users are in high-income countries which account for just 15% of the world's population (Tarjanne, 1996). The US boasts four out of ten homes owning a personal computer and one in three of these has a modem enabling the computer and telephone to be connected (see Figure 6). By the year 2000 at least half of all US homes will have two or more telecommunications lines. At present the median age of users is 32 years and dropping, 64% have college degrees and 25% have an annual income higher than $80,000. Half of Internet users have managerial or professional jobs and 31% are women. There are now more than a million Web sites for them to visit.

Figure 6: Percent of US households with a telephone, computer and Internet use

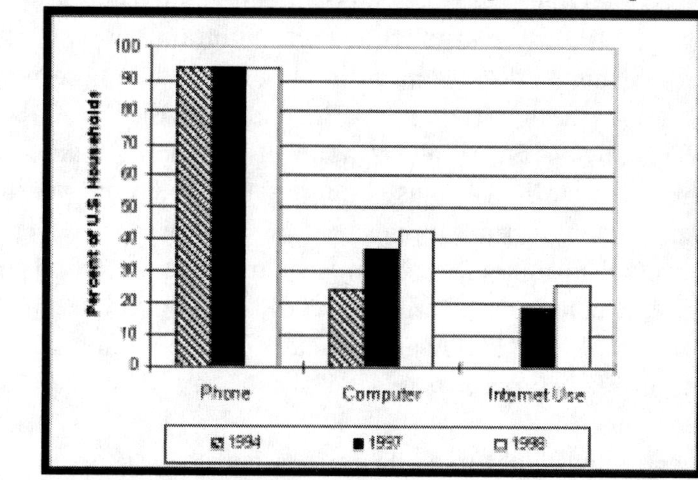

Source: Falling Through the Net, 1999

It is also useful to examine the question of the haves and have-nots from a second vantage point–a within country perspective. In fact, while the majority of the population within a developed country may qualify as "have," there is a subset of the population which does not meet the criteria. For example, the United States is considered a developed country, but the poorest 20% of households receive a smaller share of income than in almost any other developed country. Over six million homes did not have phone service in 1997 (ITU, 1998). By regions, households in Oceania (predominantly Australia and New Zealand) are the most wired, with penetration rates of over 90%. This is in contrast to Asia, where about 20% of households have a telephone, and to Africa, where the figure drops to 6% (ITU, 1998). Within country comparison by urban and rural areas also shows marked differences. Over 80% of Thailand's population still lives in rural areas, yet less than 40% of telephone lines in the country are in non-urban areas. These within country variances at best retard the overall economic growth of the respective countries.

Whether developed or developing, we also observe significant segments of the population that do not have access to the ITC infrastructure. These groups are characterized by low income, young limited education, member of a minority group, elderly, handicapped, and rural. The irony is that it is these groups that, were they to have access, would be simultaneously empowered to take steps to improve their economic well-being. It is these groups that receive huge benefits from being able to engage in job search activities, take educational classes, or access government reports online for example.

Falling Through the Net : Defining the Digital Divide. A 1999 survey of the digital divide in the US (third in a series from 1995) shows that whilst there is expanded information access, there is a persisting "digital divide" which has actually increased since the first survey (see Figures 7 and 8). The least connected are typically lower income groups, and Blacks and Hispanics. Additional geographical locations (urban, central city, and rural), age, education and household type are additional factors leading to disadvantaged groups. The following are profiles of groups that are among the "least connected," according to the 1999 data:

- **Rural Poor**-Those living in rural areas at the lowest income levels are among the least connected. Rural households earning less than $5,000 per year have the lowest telephone penetration rates (74.4%), followed by central cities (75.2%) and urban areas (76.8%). In 1994, by contrast, central city poor were the least connected. Rural households earning between $5,000-$10,000 have the lowest PC-ownership rates (7.9%) and online access rates (2.3%), followed by urban areas (10.5%, 4.4%) and central cities (11%, 4.6%).

A high-income household in an urban area is more than 20 times as likely as a rural, low-income household to have Internet access.

Figure 7: Percent of US households with a computer by income

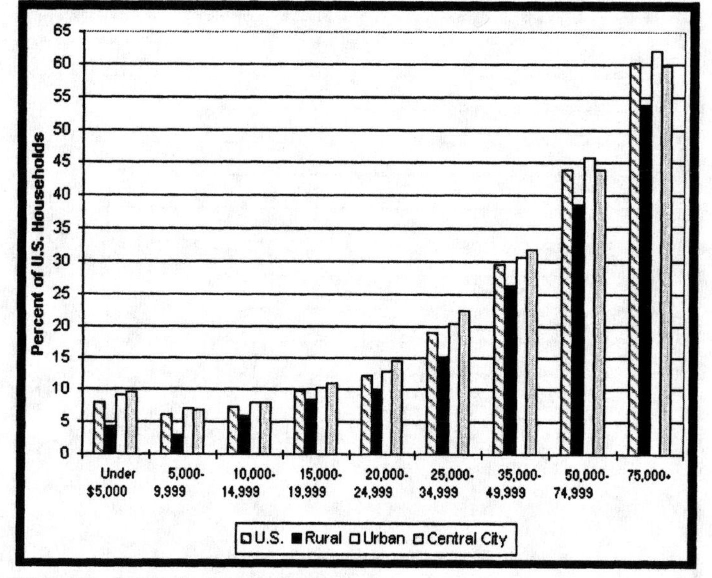

Source: Falling Through the Net, 1999

- **Rural and Central City Minorities**-"Other non-Hispanic" households, including Native Americans, Asian Americans, and Eskimos, are least likely to have telephone service in rural areas (82.8%), particularly at low incomes (64.3%). Black and Hispanic households also have low telephone rates in rural areas (83.2% and 85%), especially at low incomes (73.6% and 72.2%). As in 1994, Blacks have the lowest PC-ownership rates in rural areas (14.9%), followed by Blacks and Hispanics in central cities (17.1% and 16.2%, respectively). Online access is also the lowest for Black households in rural areas (5.5%) and central cities (5.8%), followed by Hispanic households in central cities (7.0%) and rural areas (7.3%).

To put this in simple terms: a child in a low-income White family is three times as likely to have Internet access as a child in a comparable Black family and four times as likely as children in a comparable Hispanic household.

- **Young Households**-Young households (below age 25) also appear to be particularly burdened. Young, rural, low-income households have telephone penetration rates of only 65.4%, and only 15.5% of these households are likely to own a PC. Similarly, young households with children are also less likely to have phones or PCs: Those in central cities have the lowest rates (73.4% for phones, 13.3% for PCs), followed by urban (76% for phones, 14.5% for PCs) and rural locales (79.6% for phones, 21.2% for PCs).
- **Female-Headed Households**-Single-parent, female households also lag significantly behind the national average. They trail the telephone rate for married

Figure 8: Percent of US households using the Internet by income

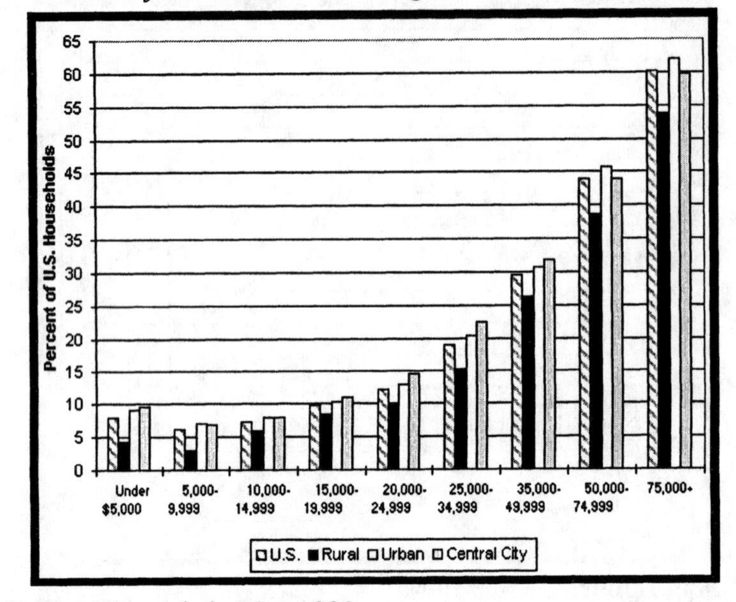

Source: Falling Through the Net, 1999

couples with children by 10 percentage points (86.3% versus 96%). They are also significantly less likely than dual-parent households to have a PC (25% versus 57.2%) or to have online access (9.2% versus 29.4%). Female-headed households in central cities are particularly unlikely to own PCs or have online access (20.2%, 6.4%), compared to dual-parent households (52%, 27.3%) or even male-headed households (28%, 11.2%) in the same areas.

The data reveal that the digital divide–the disparities in access to telephones, personal computers and the Internet across certain demographic groups–still exists and in many cases has widened significantly. The gap for computers and Internet access has generally grown larger by categories of education, income and race. This remains the chief concern as those already with access to electronic resources make rapid gains while leaving other households behind. We are witnessing the wholesale disappearance of work accessible to the urban poor. Without intervention, unemployment, poverty, and out-migration will likely increase, exacerbating the structural problems typical of rural areas (OTA, 1996).

In Australia, the picture is very similar. The report "Women's Access to Online Services," produced by the Office of the Status of Women in December 1996, states: "The Governments' focus on commerce has meant that the social consequences of becoming an 'information society' have been largely ignored. This may have been exacerbated by the apparent lack of women in decision-making positions in industry and relevant departments." The most recent data from the

Australian Bureau of Statistics (1998) estimated 262,000 users who indicated use of the Internet at home, with about 178,000 being men and 84,000 women (68%:32%). Women's representation amongst email users was even lower-at only 26%. Women over the age of 55 were extremely poorly represented. However, perhaps a more important issue is "What access opportunities are open to women who don't have a computer and modem at home?" AGB McNair estimates that in the region of only 13% of Australian women over the age of 14 have ever accessed the Internet!

Other countries' digital divides also persist; the percentages are simply higher for the have-nots. For example, Egypt's "haves/have-nots" ratio, a lower-middle income country as defined by the World Bank,[1] represents less than 8% of its 60 million plus population.

There are astonishing exceptions to the rule–one example is women farmers. The DSS CRP case studies found that women farmers "were the enthusiasts, the main drivers, while their husbands, if they had no prior computer experience, were reluctant to touch the CIN (Community Information Network)". Weather information, farming practices, health and education were all foci but, further, email was used to develop support networks, thereby reducing social and cultural isolation. Strangely it is not only those women typically identified as culturally isolated (aboriginal, non-English speaking, remote communities such as mining) but also professionally educated women whose need for professional support, continuing education and contact with like-minded peers is not adequately met.

Increasingly, education, health, legal services and social communications are moving to computer-based technology. The success of the Ipswich Global Infolinks project "SeniorNet" is another startling example. One resident said "I personally find the Internet to be a fascinating medium where any information seems available. ... [IT] opens up a whole new world for elderly people and keeps the mind active. ... There is no age limit to having a good time surfing the Net (des Artes, 1996).

The Internet is increasingly viewed as the window to the global economy. Is then the Internet the secret weapon for the have-nots? Is it for the masses? One may argue that what subsistence farmers in Afghanistan, or Korea, or Cambodia need is NOT high-tech science and complex systems, but immunizations, basic literacy, disease- and drought-resistant cereals and oilseeds, simple pumps, or deep-drop toilets. The fallacy of the pro-Internet argument is that it ignores the social and economic implications of the technology, as highlighted in this discussion.

A second argument in favor of the technology is that it will assist developing countries in leapfrogging stages in the development process. Many highly successful initiatives are taking place in developing countries to promote community-based Internet access for health (effective water sourcing, sanitation, bioengineering of crop production), educational (electronic network of schools), and other applications. The Mbendi AfroPaedia Web site (www.mbendi.co.za), the pan-European FRIENDS (Farming and Rural Information Expertise and News Dissemination Service) project, and the Mediterranean Institute of Teleactivity (IMeT) are representative of these types of initiatives (Stratte-McClure, 1999). Compelling examples demonstrate the pay-off: In rural southern Ghana, petrol stations are able to place orders with suppliers by phone when previously they could only be made by traveling to Accra; in Zimbabwe, one company generated $15 million of business by advertising on the Internet; in China, a little girl's life was saved when her doctor posted her symptoms to an Internet discussion group and received an immediate answer. Sam Pitroda, Indian government advisor, states, "IT is not a luxury but VITAL to basic activities, such as bringing food to market, preventing drought, a major source of new jobs and wealth." The conundrum is that sustainable development is an immensely complex process having its roots in educational and infrastructural building; what then is the role for the Internet in this process?

HOW CAN THE INTERNET ADDRESS THE CHALLENGE?

It is recognized that an educated population with skills and knowledge in information technology is an instrumental part of sustainable development. The irony is that while the volume of information and knowledge that is available is increasing, the percentage of the world population able to have access to and derive value from it seems to be becoming smaller. The gap between the haves and the have-nots is increasing significantly–both on a global and local basis (Novak & Hoffman, 1998). The magnitude of the challenge within countries is related to income distribution and country size. Central and Eastern European countries enjoy high teledensities in relation to their income levels because they have more even levels of income distribution than other regions (ITU, 1998). In terms of size, smaller countries are more able to reinvent themselves than countries such as China with massive populations and huge geographic expanse. Ireland and Singapore are good examples of small countries who, through aggressive national policy towards technology and education, repositioned themselves in the global market.

Is the Internet the secret weapon to bring equality to the masses-to close the gap? Fact one: Information represents power in both the political and economic spheres. Fact two: Almost every emerging society, has made it a priority to participate in the global information society bearing witness to the belief in its ability positively to affect their country's well-being. Fact three: The

Internet is the technological innovation that can provide access to the same markets and the same information within the same time frame as is the case with more developed countries. It would seem therefore that the answer to the question is "yes"-but that access is a necessary albeit insufficient solution. What then are the implications?

The traditional approach for introducing technological innovations has been through the educational system and the workplace. The problem is that a significant portion of the have-not segment does not participate in these venues. A nontraditional approach must be taken.

If access to the have-nots is to be achieved, then technological innovations, such as the Internet, need to be brought specifically to the target group and on their level. Venues where Internet' awareness, exposure, and ultimately, familiarity need to be developed. Candidate sites include communal gatherings such as the post office, hospitals, banks, and the local merchant. Furthermore, the have-nots must perceive value, an incentive to take the steps to go beyond simple awareness to becoming an actual user of the technology. The success of this effort is necessarily linked with the extent to which applications are socially and culturally appropriate and specifically address those daily life issues that concern the intended users, such as registering to vote, access to government information, access to medical information and assistance, or bus schedules.

In the local village or community where the deployment is being made, co-opting a key individual is instrumental to success. The key individual receives the benefits of training plus the respect of his or her community as the knowledge broker. They are instrumental in introducing the technology to others. Use of the systems, at least initially, will likely need to be heavily if not totally subsidized. This certainly raises the bar for many developing countries and also illustrates how country size quickly becomes a significant factor.

Egypt is an example of a country that developed a model that includes education and training, infrastructure, and IT in general and Internet access over time, all together. Moreover, it developed applications that would be culturally and socially appropriate so as to gain widespread support for the effort. But, results do not come overnight: There is a requirement for champions and long-term commitment on the behalf of the sponsor-in this case, the Egyptian government. In the late 1980s, Egypt began to deploy computer-based systems in its 27 governorates, creating Information and Decision Support Centers (IDSC) (Nidumolu et. al., 1996; Kamel, 1995, 1997; El-Sherif & El-Sawy, 1988). The effort was part of a comprehensive plan to introduce and rationalize the use of information technology in key sectors in the economy. Over time, the IDSCs have been extended into the local villages and more rural areas. There are currently 1,102 IDSC facilities. Technology

Community Centers represent the most recent efforts to introduce and rationalize the use of information technology in general, and specifically the Internet, to the general populace (Loch et al., 1999). The focus of the community centers is on children up to the age of 20. Egypt's income distribution and population demographics are typical of many developing countries. Less than 10% of the population comprises the haves subset of the population. More than 60% of its population is under the age of 30; of that segment, more than 50% is under 20 years of age. The implication is that the extent to which these segments are exposed to advanced technologies and educational and training opportunities is highly correlated with the future economic well-being of the country.

The International Telecommunication Union's (ITU) Telecommunication Development Bureau (BDT) has a program for Multipurpose Community Telecentres (MCTs) for rural and remote areas. The ITU is working in partnership with other international organizations and the private sector, installing pilot MCTs in and around a dozen countries. The operating principal of this effort is the information premise: Access to information (services) brings about improved access to the local marketplace which in turn enhances economic growth and which ultimately impacts the global competitiveness of the country. MCTs articulate the premise slightly differently, also arguing that access to information services can also help to lessen isolation and combat the problem of brain drain from rural to urban areas. Contrary to past history, where technological innovations were contributors to mass migration of people, the Internet might allow people to remain in place while making available needed information.

The project in Uganda subsidized by the BDT exemplifies the above. The MCT will be installed in Nakaseke to provide individuals with access to telephones, facsimile machines and computing facilities, including Internet access. It offers training, technical support, and professional guidance to produce electronic information reflecting local knowledge and requirements. The library is integrated into the telecentre. The MCT will provide support to teachers in the local school system through information support to the school libraries, the provision of visual resources, training for teachers in the use of computers, and distance education. The staff of the local hospital will use the facility for telemedicine applications, continuing education for health staff, and access for local health workers to medical-related resources on the Internet. Other targeted local user groups include small businesses and farmers, local councils, the women's training organization, nongovernmental organizations, and the general public (ITU, 1998, box 5.1).

SUMMARY OBSERVATIONS

True, the Internet and its associated developments, such as the World Wide Web, are a developed global phenomenon. True, the gap between the

Figure 9: Three challenges to technological innovation deployment

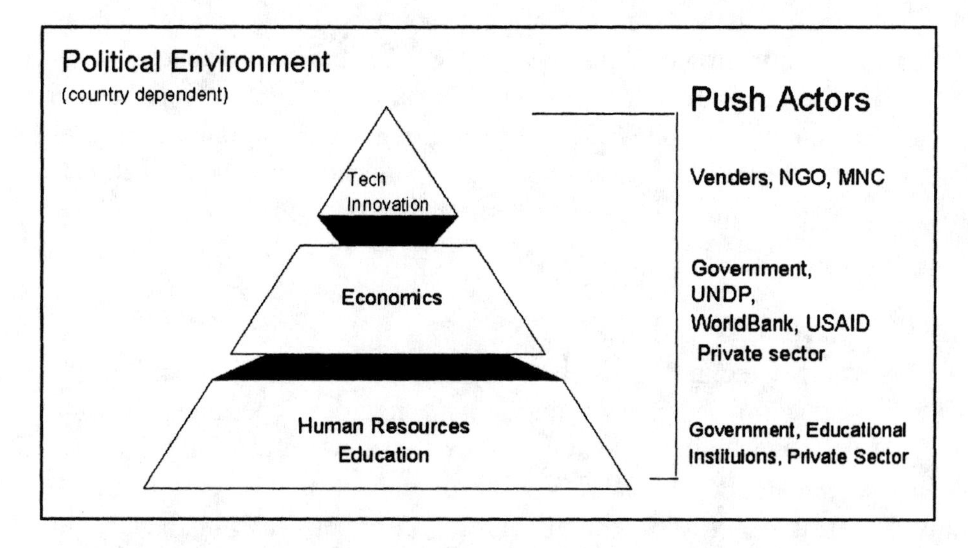

economically advantaged and disadvantaged continues to increase in both developed and developing economies. With the experiences in Egypt and Uganda as exemplars, we can make several observations that may be useful to other organizations and governmental agencies considering such initiatives or for researchers examining such initiatives.

First, existing articulated demand for technology-based services by the have-not group for such services is likely to be small to nonexistent. Hence the effort is very much characteristic of a push phenomenon (Gilbert, 1996; Gurbaxani et al., 1990). Central and local government authorities, international agencies, and leading entities from the private sector are playing, and must continue to play, key roles. Aggressive IT policy by the Singaporean government transformed Singapore within a 20-year time frame. Other countries, such as Uganda and Egypt as highlighted in this paper, are making inroads, but one must acknowledge that it is a long road to travel.

Second, there are three levels of challenges that are part of this effort. The first level is a human resource challenge. The availability of quality education and the level of literacy are both part of this challenge. The second level is the economic challenge. On an individual basis, the ability to pay for service is minimal. This places additional pressure on the providers to make the service inexpensive and widely available. On a country/governmental level, such efforts stretch the economic resources of the providing agencies. The geographic size and population distribution and size are all factors that make this level a particularly difficult challenge. The magnitude of the task for China, for example, far exceeds that of Singapore simply due to its geographic span and population demographics.

The third level is the technological innovation itself. In the case of the Internet, a base level of infrastructure must be in place to be able to deliver access to the Internet and, in turn, access to the global marketplace. All three levels are interrelated. The simple availability of the technology is insufficient; training to support its use must also be available. All levels reside in a political environment which varies from country to country. Figure 9 depicts these levels in context.

THE REAL CHALLENGE?

Information technology is generally perceived as a major facilitator for globalization, with the implication that hitherto underdeveloped regions can now gain access to worldwide resources and expertise, which will in turn lead to enhanced economic development. Globalization theorists, however, argue that it is only capital that has escaped the confines of space (Bauman, 1998; Beck, 2000). Capital has gained almost unlimited, instantaneous mobility, whereas people remain relatively immobile. One could argue that the development of global networks serves only to enhance the more developed nations and support the most dominant values, leading to increased exploitation of the less developed nations and the more disadvantaged sectors of society (Castells, 2000).

A powerful tool such as the Internet, used creatively, can serve to begin to reduce the growing and persistent gap between the haves and the have-nots but only if we begin to address the kind of problems identified in this paper.

Consider these words which come from the Cyberspace Declaration of Independence (Barlow, 1996):

- Cyberspace is a world that is both everywhere and nowhere
- A world that all may enter without privilege or prejudice accorded by race, economic power, military force, or station of birth
- A world where anyone, anywhere may express his or her beliefs
- A world where legal concepts of property, expression, identity, movement and context do not apply
- A world of no matter.

It is in our hands to make our new world matter and for it to be a cyber civilisation to be proud of. Otherwise the proud boast that: "We will create a civilisation of the Mind in Cyberspace. May it be more humane and fair than the world your governments have made before" (Barlow, 1996) will remain empty rhetoric.

ENDNOTE

[1] The World Bank has defined economic groupings of countries based on Gross National Product (GNP) per capita. Economies are currently classified based on their 1995 GNP per capita as follows: low income–economies with a GNP per capita of US$75 or less; lower-middle income–economies with a GNP per capita of more than US$ 766 and less than US$3,035; upper-middle income–economies with a GNP per capita of more than US$ 3,035 and less than $9,386; and high income–economies with a GNP per capita of more than US$8,956 (ITU, 1998).

REFERENCES

Australian Bureau of Statistics. (1998). *Household Use of Information Technology*, Australia. Cat no. 8146.0.

Barlow, J. P. (1996). *DBLP: John Perry Barlow*. Retrieved on the World Wide Web: http://www.cse.unsw.edu.au/dblp/db/indices/a-tree/b/Barlow:John_Perry.html

Bauman, Z. (1998). *Globalization: The Human Consequences*. Cambridge, Polity Press.

Beck, U. (2000). *What is Globalization?* Cambridge, Polity Press.

Castells, M. (2000). Information technology and global capitalism. In Hutton, W. and Giddens, A. (Eds.), *On the Edge: Living with Global Capitalism*. London: Jonathon Cape.

CommerceNet. (1998). *Knowledge-Internet Statistics*. Retrieved on the World Wide Web: http://www.commerce.net/research/stats/wwstats.html

Dixon, P. (1999). *Cyber Reality*. Retrieved on the World Wide Web: http://www.globalchange.com/cyberr_index.htm

El-Sherif, H. and El-Sawy, O. (1988). Issue-based decision support systems for the Cabinet of Egypt. *MIS Quarterly*, December.

Falling Through The Net: Defining the Digital Divide. (1999). Retrieved on the World Wide Web: http://www.ntia.doc.gov/ntiahome/digitaldivide/.

Gilbert, A.L. (1996). A framework for building national information infrastructure: The evolution of increased reach and range in Singapore. In Palvia, P., Palvia, S. and Roche, E. (Eds.), *Global Information Technology and Systems Management: Key Issues and Trends*, 55-76. Nashua, NH: Ivy League Publishing.

Gokalp, I. (1992). On the analysis of large technical systems. *Science, Technology & Human Values*, 17 (1), 57(22).

Gurbuxani, V., King, J. L., Kraemer, K. L., McFarlan, F. W., Raman, K. S. and Yap, C. S. (1990). Institutions in the international diffusion of information technology. In *Proceedings of 11th International Conference on Information Systems*, Copenhagen, Denmark, December, 87-98.

International Telecommunication Union. (1998). *World Telecommunication Development Report 1998*. ITU, Geneva, Switzerland.

Kamel, S. (1995). IT diffusion and socio-economic change in Egypt. *Journal of Global Information Management*, Spring, 3(2).

Kamel, S. (1997). *DSS to Support Socio-Economic Development in Egypt.* HICSS-30.

Loch, K. D., Straub, D. and Hill, C. (1999) Field interviews-Egypt, March-May.

Network Wizards. (2000). *Internet Domain Survey*. Retrieved on the World Wide Web: http://www.isc.org/ds/.

Nidomolu, S., Goodman, S. E., Vogle, D. R. and Danowitz, A. K. (1996). Information technology for the local administration support: The governorates project in Egypt. *MIS Quarterly*, June, 197-224.

Novak, T. P. and Hoffman, D. L. (1998). Bridging the digital divide: The impact of race on computer access and Internet use. *Science,* April, 17.

OECD. (1997). Information Technology Outlook. Retrieved on the World Wide Web: http://www.oecd.org/.

OECD. (1998). Electronic Commerce. Retrieved on the World Wide Web: http://www.oecd.org/subject/e_commerce/summary.htm.

Office of the Status of Women, Department of Prime Minister and Cabinet, Australia. (1996). *Regulating the Internet: Issues for Women*, December.

Office of the Status of Women, Department of Prime Minister and Cabinet, Australia. (1996). *Women's Access to Online Services, December.*

Pitroda, S. (1993). Development, democracy, and the village telephone. *Harvard Business Review*, November-December, 66(11).

Stratte-McClure, J. (1999). Trade and exchange on the Net. *International Herald Tribune*, October, 14.

Tarjanne, P. (1996). The Internet and thin information infrastructure: What is the difference? *SMPTE Journal*, October, 657-658.

The World Bank Group. (2000). Retrieved on the World Wide Web: http://www.worldbank.org/.

Zaman, R. (1999). The mobilization of Bangladesh. *International Herald Tribune*, October, 14.

Chapter 8

Method over Mayhem in Managing e-Commerce Risk

Dieter Fink
Edith Cowan University, Australia

Under the system of e-commerce, organisations leave themselves open to attack which can have catastrophic consequences. Recent well-publicised business disruptions to firms such as Northwest Airlines and Ebay have had significant business impacts. The chapter identifies the differences in risk management approaches for older information technology systems and those required for e-commerce. The benefits and the critical success factors for an e-commerce risk management methodology are identified and discussed. A literature survey revealed the existence of only two methodologies with potential suitability for e-commerce risk management. They are evaluated against the critical success factors. The chapter recommends a program of research to make risk management more dynamic and interactive particularly for the operational aspects of e-commerce.

INTRODUCTION

E-commerce is a multi-faceted business where organisations sell their products to consumers, business combine with other businesses to form virtual enterprises, and suppliers link with partners in a virtual supply chain. The driving force behind the new digital economy is the Internet and technologies, such as the World Wide Web (Web), underpinning this network of networks. The Internet can be described as a non-hierarchical, democratically structured, collaborative arrangement entered into by millions of network users.

Previously Published in *Managing Information Technology in a Global Economy*, edited by Mehdi Khosrow-Pour, Copyright © 2001, Idea Group Publishing.

Organisations practising e-commerce leave themselves open to attack and compromise which can have catastrophic consequences. New forms of crime are developing (e.g., denial of service) which have not been experienced before. These inherent insecurities require that stringent Risk Management (RM) practices are adopted. However, because of the speed with which network topologies, services and applications change and the whole interconnected nature of the business world in an e-commerce environment, RM systems must operate in a dynamic way. E-commerce is starting to create a true "just-in-time" economy and RM approaches need to reflect this.

In this chapter, we recognise the potential for mayhem in e-commerce if risk is not adequately managed. We then proceed to identify the risk areas of e-commerce and compare these with the traditional forms of IT risk. To cope with the new risk environment we outline the benefits of using RM methodologies and survey the literature to establish the existence of e-commerce oriented RM methodologies and their suitability according to criteria we established from our analysis of e-commerce risks.

POTENTIAL FOR MAYHEM

Computer systems employed to facilitate e-commerce "will become the new form of catastrophic exposure that will replace earthquakes and hurricanes as the number one form of catastrophic risk" according to Mullaney of F&D/Zurich (quoted by Hays, 2000). Gow of ACE USA holds a similar view; "If a major operation's network is intentionally compromised, it could interrupt a daily revenue in the millions" (quoted by Hays, 2000). The major compromises for e-commerce were identified as hacker's blackmail, corruption of data, disgruntled employees and unauthorised acts of system administrators.

There are changes taking place within e-commerce that have the potential for significant damage. The consequences of someone meddling with the Web site can range from mild (e.g. the introduction of a detectable virus) to catastrophic (a prolonged system outage leading to the loss of customers). Then there are risk such as out-of-date information, misinformation and defamation on the Web site, even if only there for a brief period, that can lead to lawsuits and claims for damages against the organisation. Added to this is the uncertainty of jurisdiction since the physical space has been replaced by the virtual space. "The velocity and scope of disasters that occur in cyberspace have no real boundaries" (Davis, 1999). The Internet is being called a "legal vacuum."

Negative consequences of using the Internet have been experienced by organisatons and have been well publicised. Disruptions in e-commerce have led to lawsuits, significant losses and dissatisfied customers. Table 1 shows recent e-business outages and the business impact they had.

Table 1: E-Business Outages and their Impacts

Company	Industry	Disruption	Business Impact
TD Waterhouse	Financial Services	Extended telecommunications and e-business outage	Significant financial losses
Ameritrade	Financial Services	Multiple system outage due to technical architecture weaknesses	Law suits from online traders. Significant financial losses
Northwest Airlines	Transportation	Multiple system outage	Decline in customer satisfaction
Ebay	Online Auction	Multiple system outage due to technical architecture weaknesses	Significant financial losses
Etrade	Financial Services	Multiple system outage due to technical architecture weaknesses	Law suits from online traders. Significant financial losses
CIHost	Internet Hosting	Extended system outage	Class action lawsuit from online customers. Significant financial losses. Decline in market position/ perception.

(Source: Project Management Institute - Risk Management Specific Interest Group, 2000)

CONTRASTING ITS WITH E-COMMERCE RISKS

The most common form of RM takes the approach of *risk reduction*. Under this approach, the likelihood or the impact of a threat is reduced to the level considered acceptable for the costs of implementing the security measure that reduces the threat. With *risk avoidance*, the organisation does not accept a threat and does everything within its power to prevent it from occurring. An example is the installation of an emergency power generator to eliminate the risk of being without power should the normal power source fail. Under *risk transfer*, the responsibility for the risks and costs are passed to another party, such as an insurance firm.

Risk reduction in RM is practised in many industries (e.g., in banking during loan approval) including by organisations to protect their Information Technology and Systems (ITS). They use RM to determine threats, ITS to be protected and security

Figure 1: The Traditional Processes of ITS Risk Management

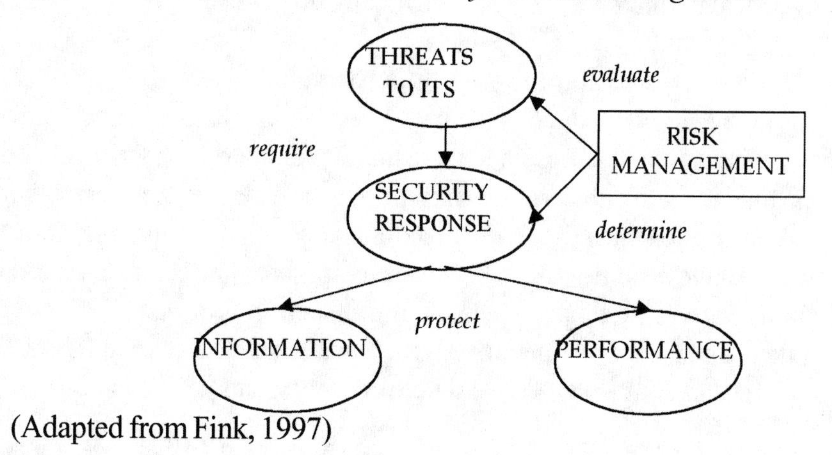

(Adapted from Fink, 1997)

measures needed to act against threats. In the case of ITS, the major assets to be protected consist of the organisation's valuable information resource and the continuity of its business operations. The dependencies between threats, security response, and resources to be protected in the form of information and performance, are shown in Figure 1.

With the introduction of e-commerce, the ITS environment has changed substantially and business is no longer conducted "as usual." While some of the risks associated with e-commerce are not new (e.g., hacking, theft of intellectual property), new implications have arisen because of the far-reaching scope of e-commerce. Old risks have been given a new twist in an e-commerce environment. To understand the new risk environment it is necessary to contrast it with that of the previous ITS environment.

Closed vs. Open Systems

With previous generations of IT, systems were less accessible and open to attack. For example, damages to stand alone systems and local area networks (LANs) are restricted inhouse. E-commerce systems, on the other hand, provide increasing levels of connectivity and accessibility to data and networks from outside the organisation. Losses dues to prolonged system outage are potentially much larger since they are noticed by the outside world and will lead to loss of business and custom.

Tangible vs. Virtual Assets

Traditional ITS environments are more tangible and were easily recognised as data processing centres. With e-commerce, information and virtual trading communities are more difficult to track. Intangible assets are important and take the form of intellectual property, information and knowledge, trademark, patents, copyright. Risks associated with virtual assets (e.g., breach of copyright) are difficult to manage and the courts have been vague about matters of online liability. The absence of physical locations complicates the RM situation.

Development vs. Operations

Systems in the past were developed in a controlled manner and released for operations after extensive testing. They were maintained during their life to add new functions and to rectify any problems. With e-commerce, the need for market responsiveness requires that system are developed and operated in a very short time. Operations have become critical because e-commerce aims at high transaction rates in order to bring down the costs of transaction processing. Greater emphasis is therefore placed on operational risk management (McEachen, 2000). Any prolonged system outage will have severe consequences.

Figure 2: The Processes of e-Commerce Risk Management

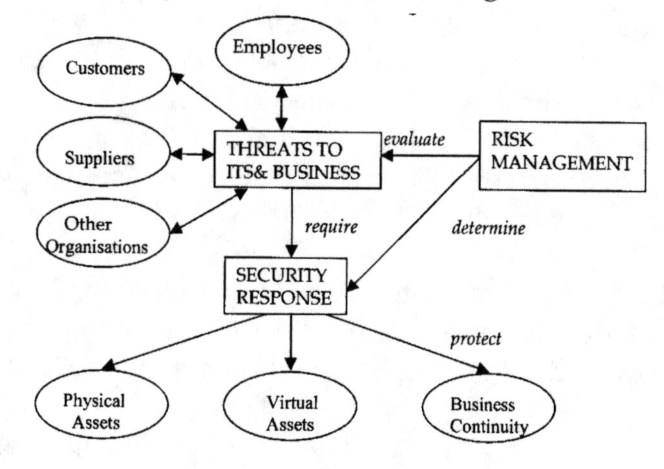

Predictability vs. Volatility

In the past, risk and security management could take place at a leisurely pace and reviews were conducted every couple of years. The RM culture for traditional ITS is unlikely to be satisfactory for the e-commerce environment. It was developed in the early days of computers when IT security could be ensured through physical measures or through the use of closed networks. With each development of an e-commerce function, new elements of risk emerge and uncertainty arises. Examples of current issues and our lack of ability to deal with them efficiently include contractual issues, legislation, taxation, liability, financial exposure and reporting (see McDonald, 2000) to mention a few.

Compared to the RM processes of older ITS (see Figure 1), those for e-commerce have become more complex and greater inter-dependencies have to be considered. Furthermore, the nature of assets to be protected has changed and business continuity has become critical. The changes are reflected in Figure 2.

RM METHODOLOGIES AND BENEFITS

With the increased complexity of e-commerce, compared to former generations of ITS, the need for a methodological approach to RM has increased. A methodology can be described as a "A logical and systematic method of identifying, analysing, evaluating, treating, monitoring and communicating risks associated with any activity, function or process in a way that enables organisations to minimise losses and maximise opportunities" (Standards Australia, 1999). For e-commerce it is essential that RM methodologies build on the generic fundamentals of RM but develop approaches that meet requirements of the new networked, virtual environment. These specialised requirements are identified and outlined in the following section of the chapter.

The generic approach to RM is reflected in standards that have been developed for security management over time. In the UK, BS7799 provides a code of practice for information security management. A key component of these standards is the requirement that risk assessment is carried out. The standard allows compliant companies to publicly demonstrate that they can safeguard the confidentiality, integrity and availability of their customers' information. Standards have the potential to become the de facto "seal of approval" in the world of e-commerce.

Similar standards exist in other countries. In Australia, AS/NZS4444 is based on BS7799 and establishes a code of practice for selecting information security controls (AS/NZS4444.1) and specifying an information security management system (AS/NZS4444.2). As does BS7799, both parts of AS/NZ4444 require that a risk assessment process is used as the basis for selecting controls (treating risks). If an organisation wants to trade securely over the Internet it should ensure that both itself and its partners have this accreditation.

While it may appear that there are overheads associated with the adoption of a RM methodology, there are a number of significant benefits associated with its use.

Stakeholders Participation

Information security is often seen solely as the domain of IT professionals with information security being viewed from a technological aspects, i.e., implementing technological measures such as firewalls and encryption. Kelly (1999) refers to this practice as "point solutions" – quick fixes that can do more harm than good. Management and users appear rarely involved in the process. With the increasing integration of ITS in all business activities, as occurs with e-commerce, stakeholders from all areas of the organisation need to be involved. A methodology can provide the framework that will ensure representatives from all business operations participate in risk management.

Holistic

"Many security problems are caused by all too human misperceptions of where dangers actually lie and the ability of particular measures to avoid them" (Brewer, 1999). The levels of security understanding of security threats, exposures, safeguards, practices and priorities among information users and solution providers varies widely. Risk management therefore needs to be approached in a systematic manner so that all perceptions are included and evaluated. A RM methodology provides an approach in which all perceptions, covering all aspects of e-commerce, are captured, analysed and communicated.

Competitive Advantage

RM and security activities are seen as burdensome practices that create additional work for already stretched resources. They are often perceived to be only needed when the organisation is under attack or special circumstances arise. This negative perception needs to be reversed and instead a perception that RM can become a competitive advantage to the firm needs to be created. Security should be seen as an enabler since, by operating safely, the organisation can take more risks than its competitors. RM has the obvious advantage of preventing expensive system outages.

CSFS FOR E-COMMERCE RM METHODOLOGIES

RM for e-commerce is different to RM for older ITS because of the differences in their risk profiles (see earlier discussion). An e-commerce RM methodology should nevertheless build on the fundamentals of good security management principles. These can be summarised as follows. "Generally, information security risk management methods and techniques are applied to complete information systems and facilities, but can also be applied to individual system components or services where this is practicable, realistic and helpful. It is an iterative process consisting of steps, which, when undertake in sequence enable continual improvement in decision making" (Standards Australia, 1999). Based on these principles, and the need to consider technology and business risks, a number of Critical Success Factors (CSFs) for an e-commerce methodology can be identified as follows.

Is the Methodology Effective for E-Commerce?

Comprehensive

It must cover technical and business scenarios that are part of the various types of e-commerce (business-to-business, business-to-consumer), the phases of the e-commerce development (from planning to implementation) and the life cycle of operations (from ordering to supply and payments).

Inclusive

The methodology must cover all assets, vulnerabilities and threats. They include technology and business assets, real and virtual, by themselves as well as their interactions. With e-commerce, technology and business are integrated and work synergistically together to achieve maximum impact.

Flexible

It must offer a variety of techniques that can be applied across some or all phases of the methodology. Traditionally, an organisation's assets and policies may have been static but threats in e-commerce are mobile and mutable. Threat levels can rise and fall during a single day's operations as new sites come online or a new virus emerges.

Pro-active

The methodology must be flexible and promote pro-activity to anticipate changes in the e-commerce environment. It should encourage pro-active behaviour that uses RM to gain competitive advantages. To be effective, the methodology should provide a simple approach suitable for an inherently dynamic environment..

Relevant

RM should lead to the identification and application of security measures relevant to e-commerce. Security techniques for e-commerce include the installation of firewalls to separate the 'untrusted' outside networks from trusted internal ones, the use of digital signatures and certificates, encryption, etc.

Will the Methodology Provide a Competitive Advantage?

Credibility

The RM methodology should comply with an accepted standard such as BS7799, and should be supported by a credible vendor who is able to provide training, support, documentation, updates, consulting and implementation services. Furthermore, it should offer accreditation by reputable associations and industry bodies.

Value

The cost of the use of the methodology should be covered by the benefits realised from its use. In the real world, resources are limited and decisions about trade-offs have to be constantly made. The RM methodology should be easy to justify in terms of the advantages it provides.

Integration

With e-commerce it is imperative that decisions are made based on both business and technological considerations. Risks in the technological domain interact with those in the business domain and a RM methodology should cover both types of risk.

Can the Methodology Be Implemented Readily?

Systematic

The processes of RM should be structured and systematic to encourage organic management behaviour, transparency and open communications. Guidelines should be available for processes to be followed and the deliverables to be produced for each activity and phase.

Adaptable

The RM methodology must be able to be customised to the existing ITS environment, organisational culture and resource constraints with the objective of making an uncertain environment more certain.

Timely

The methodology must be carried out speedily because of the rapid changes that can occur for e-commerce. It must therefore define procedures, deliverables and timelines that can be applied to small as well as major changes. A speedy implementation of the methodology is essential so that the fluency and flexibility of e-commerce is ensured.

Tracking

With increased operational risk emerged the need to measure and monitor risk factors through risk indicators and metrics. A risk management methodology should provide the system for this and ideally produce dollar-at-risk-type figures.

Sponsorship

It is generally accepted that projects fail if not supported by senior management. They should therefore be an integral part of risk and security solutions. The e-commerce sponsor should have an active involvement in implementing the RM methodology and be satisfied with risk outcomes before releasing further funding for the e-commerce project.

SURVEY OF E-COMMERCE RM METHODOLOGIES

A search of the Web and our university's online library database revealed many papers discussing the nature of e-commerce risk but very few that addressed the requirements and practices of RM for these risks. The literature on e-commerce risk appeared to be dominated by opinions held by the insurance and finance industries. These industries have long used RM to assess the risk of their clients and are developing an interest in providing insurance coverage to organisations with e-commerce activities.

In the absence of suitable methodologies on how to manage the risks associated with e-commerce, it is not surprising that the increased levels of concern have given rise to reactive approaches in the form of protective insurance (Hays, 2000). Companies with a high volume of e-commerce are under pressure from shareholders to ensure that their operations stay afloat and, according to Mullaney of F&D/Zurich, will have to "buy a risk transfer product or install a solution" (quoted by Hays, 2000).

At present, however, insurance does not provide a complete solution to organisations concerned with e-commerce risk. First, "Traditional insurance companies insure property losses and liabilities but not those that arise from e-commerce" (Davis, 1999). Second, "Currently there are fewer than 10 agencies that offer e-business insurance" (Davis, 1999). Organisations therefore have to look to other solutions particularly RM methodologies to manage e-commerce risk. Our literature search, however, revealed only two methodologies that were associated with the use of e-commerce. They are outlined below.

Active Risk Management (ARM)

The ARM approach appears to be supported by the Project Management Institute Risk Management Specific Group since it was outlined in its March 2000 newsletter (www.risksig.com/signews/Rmnews0003.pdf) whose theme was 'Continuity in a Virtual World'. In the newsletter, Parker (2000) outlines the ARM methodology which he describes as "a discipline and environment of proactive decisions and action to assess continuously what can go wrong, determine what risks are important, their impact, and implement a strategy to deal with those risks."

ARM's primary aim is to manage risk in a software project, and its proactive philosophy and methodical approach makes it potentially useful to e-commerce systems. Parker (2000) emphasises that ARM is an ongoing process and not a static project task. It includes three major phases- risk identification, risk analysis, and risk control. Phase 1 (risk identification) is accompanied by a questionnaire that narrows the focus on particular aspects of risks and assists in identifying risks during a project in the areas of requirements determination, design, code and unit test, integration and test, and communications and team motivation.

In Phase 2 (risk analysis), risk probability, risk impact, risk exposure and risk consequences are 'quantified' as best as possible. Where quantification is not possible, qualitative categorisation is applied. For example, an impact/probability matrix determines risk exposure as very high, high, medium, low and very low. Phase 3 (risk control) defines steps to avoid, mitigate or accept for risks identified in the previous phase.

CCTA RISK ANALYSIS AND MANAGEMENT METHODOLOGY (CRAMM)

CRAMM is a formalised, structured security risk analysis and management methodology developed by the British government. It is regarded by many as the

Table 2: Evaluation of e-Commerce RM methodologies

CSFs	ARM	CRAMM
Effectiveness		
Comprehensive	√	√
Inclusive	√	√
Flexible	√	√
Pro-active	√	√
Relevant	?	√
Competitive advantage		
Credibility	√	√
Value	?	?
Integration	√	√
Implementation		
Systematic	√	√
Adaptable	√	√
Timely	√	√
Tracking	?	?
Sponsorship	√	√

de-facto standard for risk analysis and management. It has been described as "an essential first step towards BS7799, the standard for information security management" (Logica) and is consistent with the European IT Security Evaluation Criteria (CRAMM User Group). Even though CRAMM was developed in the early days of computers, it has undergone changes with newer versions reflecting the requirements of new ITS environments. Version 3.0, released in 1997, is implemented as an interactive software tool for identifying the security requirements (Gamma).

A CRAMM review may be undertaken during systems development or retrospectively for completed systems and proceeds through three stages: identification and valuation of assets, quantification of likely threats and known vulnerabilities, and generation of countermeasures. CRAMM includes a countermeasures database which provides three levels of detail: security objectives, detailed countermeasures descriptions and implementation examples.

At the completion of each stage, formats can be extracted which can be in redefined and customised. CRAMM enables the building a model encapsulating asset interdependence and information on which parts of the system support which business processes. This provides insight into operational characteristics which is critical in an e-commerce environment.

Based on the information we were able to gather, we attempted to evaluate the two methodologies against criteria for a suitable e-commerce RM methodology we had established (see earlier section). Table 2 reflects our assessment.

CONCLUSIONS AND FUTURE RESEARCH

Our preliminary findings indicated that both methodologies have strong potential to provide the necessary guidelines to assist an organisation reduce the risks of e-commerce. CRAMM appears to have an advantage over ARM because it offers a countermeasure database. We are not familiar with the security and control measures included in this database but assume that they would have relevance to e-commerce because the latest version of CRAMM had been released fairly recently. As seen in Table 2, we were not able to draw a conclusion as to the tracking capabilities of the methodologies or their values because of the lack of detailed information.

Our research is of an exploratory nature and established the scarcity of RM methodologies designed for e-commerce. As e-commerce becomes mainstream, the demand for such products will increase. Our paper identified the benefits and the essential criteria for evaluating RM products that may emerge in response to this expected demand. This should provide useful insight and information to managers responsible for e-commerce risk and security.

There are a number of opportunities for researchers to support the development of e-commerce RM methodologies. As identified earlier in the paper, e-commerce requires emphasis to be placed on operational risk management. To manage operational risk requires definition and identification of key factors, such a data collection and communicatons, and key risk indicators and risk measurement metrics. A risk management methodology should provide the system for this and ideally produce a dollar-at-risk-type figure. However as Deborah Williams of Meridien Research (quoted in McEachern, 2000) points out, few vendor packages in the RM domain currently provide such output.

In future, the speed with which new networking and operating architectures and applications will emerge could outstrip the ability of management to keep up. If they can't, it will severely limit the organisation's ability to exploit new business possibilities. New generations of methodologies that automate the process of RM will be demanded. Researchers should assist in the development of RM methodologies that are able to scan a network and produce real-time analyses of vulnerabilities and effective countermeasures in forms that senior executives can understand.

REFERENCES

Brewer, D. (1999). Keeping Virtual Worlds open for Business, *Telecommunications*, 33(9), 65-66.

CRAMM User Group. What is CRAMM? (online) www.crammusergroup.org.uk.

Davies, R. (1999). Taking the Risk out of e-Commerce, *Techwatch*, October, 45-46.

Fink, D. (1997). *Information Technology Security - Managing Challenges and Creating Opportunities*, CCH Australia, Sydney.

Gamma. CRAMM, (online) www.gammassl.co.uk/topics/hot5.html.

Hays, D. (2000). Insurers Cover Hackers' Threat to E-commerce, *National Underwriter*, 20-25.

Kelly, B.J. (1999). Preserve, Protect, and Defend, *The Journal of Business Strategy*, 20(5), 22-25.

Logica. CRAMM, IT Security Risk Assessment and Management, (online) www.logica.com/offerings/CRAMM.html.

McEachern, C. (2000). Infinity launches e-Commerce Strategy, Internet Risk Applications, *Wall Street & Technology*, May, 62.

Parker, G. (2000). Active Risk Management, (online) www.risksig.com/signews/RMnews0003.pdf.

Project Management Institute – Risk Management Specific Interest Group. Risk Management Newsletter, (online) www.risksig.com/signews/RMnews0003.pdf.

Standards Australia (1999). (online) www.standards.com.au/Catalogue/Script/Details.asp?DocN=stds000023835.

Chapter 9

Why Do We Do It If We Know It's Wrong? A Structural Model of Software Piracy

Darryl A. Seale
University of Nevada, Las Vegas, USA

This study examines predictors of software piracy, a practice estimated to cost the software industry nearly $11 billion in lost revenue annually. Correlates with software piracy were explored using responses from a university wide survey (n=589). Forty-four percent of university employees reported having copies of pirated software (mean = 5.0 programs), while 31 percent said they have made unauthorized copies (mean = 4.2 programs). A structural model, based in part on the theory of planned behavior (Ajzen, 1985) and the theory of reasoned action as applied to moral behavior (Vallerand, Pelletier, Cuerrier, Cuerrier & Mongeau, 1992), was developed which suggests that social norms, expertise required, gender, and computer usage (both home and at work) all have direct effects on self-reported piracy. In addition, ease of theft, people's sense of the proportional value of software, and various other demographic factors were found to affect piracy indirectly. Theoretical as well as practical implications for the design and marketing of software are discussed.

INTRODUCTION

If dollar estimates are correct, software piracy rivals organized crime as one of our nation's most costly offenses. Although scholars are far from agreement on the level of legal protection that should be afforded software and other forms of intellectual property (Nelson, 1995) and engage in considerable debate regarding the actual costs of software piracy (Masland, 2000), most researchers agree that piracy is widespread. Industry surveys estimate that for every legitimate copy of software, there are between two and ten illegal copies (James, 2000; Conner & Rumelt, 1991). In some studies, over half of those surveyed admitted that they had made unauthorized copies of computer software. Even in the more conservative business arena, estimates suggest that in the US 25% of all installed applications are pirated. The Business Software Alliance (1999) estimates that, worldwide, the industry is losing nearly $11 billion annually in lost revenue. In the US alone, lost sales are estimated at $2.8 billion, plus a loss of over 100,000 jobs, amounting to $4.5 billion in wages and $991 million in tax revenues.

Beyond the economic impact, studying software piracy is important for other reasons. First, it may help us better understand how social norms and moral standards develop for new technologies, especially technologies involving intellectual property issues. Second, research on software piracy may expand the important philosophical debate on intellectual property. A central controversy in this debate is that many of the owner's rights commonly associated with tangible property are not violated when intellectual property is copied or used by others. Further, many philosophers and economists contend that intellectual property rights should not be protected by law (Davidson, 1989), arguing the such protection is anticompetitive, monopolistic, and can stifle creativity and progress (Abbott, 1990; Cooper-Dreyfuss, 1989; Davidson, 1989; Samuelson, 1989; Wells-Branscomb, 1990). The many proponents of stronger copyright and patent protection argue that property rights should be strictly enforced, claiming that piracy is an insult to hardworking inventors and essential to foster innovation in one of the largest value-added industries in the world (Schuler, 1998). A final reason for studying piracy behavior, and an important theme of this book, is that understanding society's norms and values regarding piracy adds to our understanding of social responsibility in the information age, which has widespread implications for design and marketing in the software industry.

THEORY AND MODEL DEVELOPMENT

Software piracy has been investigated from varied disciplinary perspectives, including: (1) economics (Gopal & Sanders, 1998; Bologna, 1982); (2) those that attempt to deter or detect would-be offenders (Holsing & Yen, 1999; Jackson, 1999; Sacco & Zureik, 1990); (3) as a risk-taking phenomenon (Parker, 1976); (4) or simply by the failure of society's morals to keep up with the growth in technology (Johnson, 1985). Much of the empirical research on software piracy has focused on ethical and legal aspects (Im & Koen, 1990) while a few studies have dealt with the social costs (Briggins, 1998; Conner & Rumlet, 1991; Mason, 1990), or attitudes (Reid, Thompson & Logsdon, 1992; Sacco & Zureik, 1990; O'Brien & Solomon, 1991; Taylor & Shim, 1993). While these studies offer some insight into various motivations to pirate, a more encompassing and plausible model for software piracy has yet to emerge. An important aim of this chapter is to develop a model that both predicts and explains incidents of software piracy.

We begin by examining previous studies that report correlates of software piracy behavior and computer use. Recognizing that the act of piracy may hinge on moral, ethical or attitudinal concerns, we turn next to several popular theories of reasoned action for guidance. Finally, to ensure that the model generalizes to a broader class of items deemed intellectual property, we examine several important aspects that distinguish tangible from intellectual property. The model that emerges integrates previous research on correlates of piracy behavior and a rational action perspective on moral behavior with several defining characteristics of software and other forms of intellectual property.

Computer Usage and Demographic Factors

Earlier research aimed at understanding software piracy approached this behavior as a dimension of computer use or, more specifically, misuse. Sacco and Zureik (1990) found that piracy was the most frequently reported misuse of computers, with 62% of respondents reporting that they had made illegal copies of software. Respondents also reported that they believed a great deal of illegal copying was going on, and that the likelihood of detection (getting caught) was very low. Previous research that examined personal and/or demographic factors has yielded mixed results regarding the relationship between gender and software piracy. One study found a significant relationship (O'Brien & Solomon, 1991), while another study (Sacco & Zureik, 1990) did not. The effects of age and computer use have also yielded divergent levels of software piracy across studies. In addition, studies have found that software pirates are generally bright, eager, motivated, and well qualified (Parker, 1976). These are the same characteristics we value in people, and beg the question: Are we trying to predict software piracy or good citizenship? Thus, a more definitive relationship between demographic variables and software piracy remains an empirical question.

Morality, Ethics and Reasoned Action

Several studies have examined software piracy with a moral or ethical focus. Im and Van Epps (1991) see piracy as yet another sign of the moral decay in corporate America. To combat the problem, they offer three prescriptions centering on educating employees concerning what is acceptable and unacceptable behavior. Swinyard, Rinne and Kau (1990) argue that many people weigh the outcomes or benefits of illegal copying more than the legal concerns of getting caught. Their results also indicate that morality judgments likely differ by culture or national origin, adding yet another dimension to be addressed in our understanding of software piracy. Although the two studies mentioned above offer somewhat different perspectives on piracy, both point toward morality and ethics and important considerations for any theory of software piracy.

Although important, theories of morality and ethics are not sufficient for developing a predictive model of piracy behavior. Here we turn to the theory of reasoned action (Fishbein & Ajzen, 1975) and two important extensions: the theory of planned behavior (Ajzen, 1985) and the theory of reasoned action as applied to moral behavior (Vallerand et al., 1992). A central feature in the theory of reasoned action is the individual's intention to perform a given behavior. Intentions capture the sum of an individual's motivational influences; they are indications of planned effort or of how hard one is willing to work to perform a behavior. Accordingly, there are two main determinants of intention: a personal or "attitudinal" factor and a social or "normative" factor. Attitude, in this context, refers to the favorable or unfavorable evaluation of behavior. It is a function of the salient beliefs one holds regarding the perceived consequences of performing a behavior and the evaluation of these consequences. Social norms consist of a person's perception of what important referent groups think he or she should do. These subjective norms are often determined by normative belief structures and motivations to comply with the behavior. Therefore, when attitudes and subjective norms coincide there is a greater intention to perform the behavior.

The hypothetical independence of attitudinal and normative factors has been seriously challenged by research showing significant correlations between these constructs (Miniard & Cohen, 1981; Ryan, 1982; Shephard & O'Keefe, 1984). These findings are particularly interesting because they suggest that either a common antecedent exists, or one's normative beliefs causally affect one's attitudes. Causality questions take on added importance when the behavior in question has moral implications.

Vallerand et al. (1992) extended the basic rational action perspective by incorporating moral behavior. They contend that a person's normative beliefs, i.e., what important others may view as appropriate behavior, are common determinants of an individual's attitudes and subjective norms. Therefore, when confronted with a moral situation, such as software piracy, individuals decide on the basis of their attitudes toward the behavior (determined in part by the probabilities and consequences of getting caught) and their perceptions of what important others (e.g., parents, other relatives, friends, professors) think is appropriate. This view is similar to differential association, a "learning" theory of deviant/criminal behavior, that suggests we adopt attitudes favorable or unfavorable to deviance based partially upon the acceptance of the attitudes and behaviors of esteemed others with whom we interact or observe (Sutherland, 1947; Akers 1994).[1]

A somewhat different configuration of measures leading to behavioral intentions is derived from a theory of planned behavior (Ajzen, 1985), however, they retain, largely, the same meaning outlined above. Attitudes, subjective norms, and perceived behavioral control are proposed as theoretically independent determinants of intended behavior. The latter concept represents the perceived difficulty of performing the behavior based upon past experience and anticipated barriers or hurdles (e.g., time, skills, cooperation of others; see Ajzen, 1985, for a more complete discussion). The relative importance of these concepts in predicting behavioral intention is expected to vary across behaviors and populations. However, generally, as attitudes and subjective norms become more favorable and the level of perceived behavioral control increases, the intention to perform a particular act should become more likely.

Modeling Conceptual Differences in Software and Intellectual Property

The theories and extensions outlined above provide an important framework from which to examine software piracy. However, due to important conceptual differences, a theory of software piracy or a more general theory covering intellectual property may differ in certain respects to reasoned action theories. Consistent with reasoned action theories both attitudinal and social factors are expected to be important determinants of piracy. People make bootlegged copies of software, music, or VHS tapes with little regard for the legality of copyrights or patents. The awareness that others are doing it can help establish a social norm that software piracy is acceptable. Among computer users, for instance, general agreement that software is overpriced or that copying is appropriate when the original software was purchased may lead to widespread approval of software "sharing" without any remorse.

If, as past research suggests, a moral component of software piracy exists, then personal attitudes and social norms are likely to be determined by common antecedents. Thus, a model of software piracy should include certain exogenous factors, such as age, income and employment, that serve to shape one's normative beliefs. These normative beliefs, in turn, are important determinants of both attitudes and subjective norms. Yet, little guidance exists regarding how such a perspective might be empirically modeled.

Additionally, perceived behavioral control is expected to play an important role within this integrated perspective. Software piracy requires certain skills and expertise, as well as opportunity. If the required abilities are beyond an individual's perceived control, software piracy is not likely to emerge. Thus, level of expertise (perceived behavioral control) is expected to have a direct effect on piracy behavior. Similarly, logistical concerns such as likelihood of being detected, ease of piracy, and opportunity should also be related to software piracy.

Notwithstanding the considerable agreement outlined above, there are several reasons to expect differences between reasoned action theories and a general theory of software piracy. Intellectual property has unique qualities that differ substantially from tangible property (Cooper-Dreyfuss, 1989). For example, it is harder to maintain exclusive use over intellectual property than it is to control tangible property. Theft of tangible property, say, a car, deprives the owner of its use as well as the right to sell it, borrow it, or trade it in on a newer model. Intellectual property, on the other hand, is not consumed by its use and can be many places at once; therefore, the possession or use by one person does not preclude others from using it (Cooper-Dreyfuss, 1989; Hettinger, 1989). Such differences outline the *nonexclusive* nature of intellectual property, which may also increase the likelihood of piracy.

Another aspect common to intellectual property is that its retail price often does not reflect its production cost. Potential purchasers may sense an inherent unfairness in the price of software; that is, they may recognize the low marginal cost to the manufacturer of producing one more copy compared to the high retail purchase price. Perceptions of price and value are important to consumers (Zeithaml, 1988), and our sense of *proportional value* requires that the price of our purchase be reflected in its cost (Hettinger, 1989). Thus, people who report an inherent unfairness in the price or proportional value of software may be more likely to engage in piracy. Moreover, the violation of one's proportional expectations may also have consequences in shaping one's attitudes or subjective norms.

The foregoing discussion provides a partial list of the items that one might expect to affect software piracy. Computer use, demographic factors or other personal attributes, social norms, proportional value, the nonexclusive nature of software, expertise, and ease of piracy were identified as likely predictors. The temporal order of several of these items, however, remains unclear. If, as past research suggests, a moral component of software piracy exists, then judgments concerning proportional value, expertise and social norms are likely to be determined by common antecedents, like personal attributes and demographics. We would also expect attitudes, like one's sense of proportional value, might have consequences in shaping social norms. Thus, regarding temporal order, proportional value is seen as an antecedent to social norms. Similarly, the level of expertise required to pirate software might be impacted by one's impression of the difficulty of pirating software. Thus, we might expect ease of theft to come before expertise.

METHOD

Thus far we have identified likely predictors of software piracy and suggested a possible temporal order for many of the variables. To test these conjectures, a questionnaire was designed that captures computer use and demographics using several single-item measures, and proportionality, ease of theft, social norms and expertise required with four multi-item measures. After appropriate pretesting, the survey was conducted and the data analysis submitted to structural equations modeling (specifically, LISREL). This technique has been used in a variety of research domains in the behavioral and social sciences and is well suited to test relationships between single and/or multi-item measures where temporal order remains important. We begin by describing the sample, survey design, and procedure.

Sample

The study was conducted at a large southwestern university. The sampling frame was a mailing list of 9550 names purchased from the university that included everyone on the university payroll, from high level administrators to full-time gardeners and part-time graduate students. In total, 1910 surveys were distributed to a random sample of employees. Of the total distributed, 589 surveys were returned (gross response rate of 31%). The study excluded 66 respondents who did not report microcomputer use (17 respondents did not answer questions concerning their microcomputer use, and 49 respondents reported they did not have access to a microcomputer either at home or at work). Thus, the final sample included 523 returned surveys from university employees who reported some microcomputer use.

The respondents represent a wide cross section of employees. Approximately 42% were classified staff, 20% faculty, 18% professional staff, 12% graduate students, and 8% administration. Men and women were equally represented (49% versus 51%, respectively), with an average age of 39.7 years old and a median education level of a college degree.

Survey Design

The survey was divided into three sections. The first section contained questions addressing general aspects of computer use: frequency of computer use, types of computer use and software applications, access to personal computers, and purchases of software. This section also contained three questions concerning self-reported piracy behavior, as well as a question concerning perceptions about the frequency of piracy among computer owners.

The second portion of the survey addressed attitudes and perceptions regarding unauthorized copying of software. Multi-item scales were developed to measure proportionality, social norms, expertise required, and ease of piracy. Response categories were 7-point Likert scales where 1 is "strongly disagree" and 7 is "strongly agree."

Nonexclusivity (nonexclusive nature) was measured by comparing responses to two questions: 1) It's alright to take home up to $25 worth of company office supplies, and 2) It's alright to copy company software that costs as much as $25. Although the dollar amounts were chosen arbitrarily, the distinction between tangible and intangible property is clear. Further, the questions were spaced to impede deliberate comparison. If the arithmetic difference between the two responses was positive, the respondent felt it was more acceptable to take home an unauthorized copy of company software than it was to take home company office supplies that have similar value. If the arithmetic difference was zero, this is an indication that the respondent sees no difference between these actions.

The third section of the survey contained the standard demographic questions of age, education, gender, religious affiliation, employment status, and income.

Procedure

The survey was pretested among select faculty, staff and graduate students. After some modifications, the instrument was distributed through campus mail to a systematic random sample of employees. As software piracy is a controversial topic, underreporting of piracy was a concern. To address this concern, the cover letter assured complete anonymity. To enhance response rate, everyone who received the initial questionnaire was sent a follow-up postcard two weeks later which asked them to respond if they had not already done so.

RESULTS

Descriptive Statistics

Of the employees surveyed, 44% reported that they have received unauthorized copies of software from friends or relatives. When asked "how many copies," the mean response was 5.0 programs (sd = 1.30). Thirty-one percent of those surveyed said that they have made unauthorized copies. When asked "how many copies," the mean response was 4.2 programs (sd = 1.32). When asked to estimate what percent of microcomputer owners have unauthorized copies, the mean response was 66% (sd = 8.30). Several questions were asked regarding the respondent's level of computer experience. Only 49% of those surveyed said they had taken two or more formal computer courses. Yet, 88% reported that they use a PC at work. Of these, 83% reported more than two years experience. Word processing (91%), spreadsheet (46%), and email (41%) were the most frequently

Table 1: Means, standard deviations, skewness and kurtosisof measurement model indicators

	Indicator Variable	Mean	SD	Skewness	Kurtosis	
1.	Overpriced: (Software is)	2.78	(1.46)	0.480	-0.336	R*
2.	Profitable: (companies developing software are)	2.63	(1.37)	0.600	-0.117	R*
3.	Value: (compared to other goods software is a)	3.94	(1.46)	-0.091	-0.221	
4.	Fairly priced; (software is)	3.35	(1.44)	0.109	-0.334	
5.	Obtained: (PC owners can easily obtain unauthorized software)	5.06	(1.49)	-0.540	-0.191	
6.	Installed: (Unauth. Software is easily)	4.83	(1.48)	-0.446	-0.010	
7.	Copies Made: (It is easy to copy unauthorized software)	5.41	(1.42)	-0.872	0.594	
8	Needs: (It's alright to copy software for bus./prof.)	2.69	(1.75)	0.938	-0.088	
9.	Workcopy: (It's alright for employees to copy company software costing $25)	2.26	(1.59)	1.301	0.935	
10.	Copy: (It's alright to copy microcomputer software)	2.89	(1.73)	0.707	-0.341	
11.	Permission: (It's wrong to copy software without)	2.64	(1.77)	0.988	0.012	R*
12.	Borrow: (Copying software is more like borrowing than theft)	2.73	(1.70)	0.751	-0.383	
13.	Friends: (It's wrong to copy software obtained from a friend)	3.25	(1.92)	0.492	-0.883	
14.	Personal: (It's alright to copy software for personal use)	2.99	(1.83)	0.583	-0.710	
15.	Make: (After purchasing software it is okay to copy it for friends)	2.90	(1.75)	0.629	-0.575	
16.	School/work: (It's wrong to copy software obtained from school or work)	2.89	(1.83)	0.730	-0.502	R*
17.	Computer Knowledge: (People would purchase if they did not have knowledge to copy)	3.85	(1.96)	0.149	-1.180	
18.	Computer Skills: (People making copies possess special skill)	2.98	(1.63)	0.622	-0.355	

NOTE: Items scored 1=strongly disagree, 7=strongly agree. Items marked with an (R*) were reverse scored to coincide with the remaining items in the factor.

cited applications. Using a PC at home was reported by 58% of the respondents. Word processing (95%), games (48%), and spreadsheets (44%) were the most common home uses. When asked if they have ever purchased PC software, 62% said yes. Of those that own or have access to a PC at home, 86% reported purchasing software.

Indicators of the Measurement Model

The 18 indicators presented in Table 1 were employed to operationalize a measurement model of software piracy perceptions that ranged from computer knowledge required to social norms surrounding the unauthorized use of software. As outlined in the table, the survey questions crossed employment, social, and personal boundaries where software piracy may occur. Indicator variables were collected using seven-point Likert scales, with four variables (overpriced, profitable, permission and obtained) reverse scored. For a more detailed discussion of the indicators, including a discussion of skewness and kurtosis, see the appendix.

Empirical Assessment of the Measurement Model

In exploratory analyses not shown here, several additional factors of deterrence, or detectability, and nonexclusivity were included. However, these factors were found to have no empirical basis. Further, the single indicators of deterrence and nonexclusivity were found to be unrelated to software piracy and are not included in the structural model. Several indicators were also eliminated from the

Figure 1: Measurement model: Software piracy perceptions

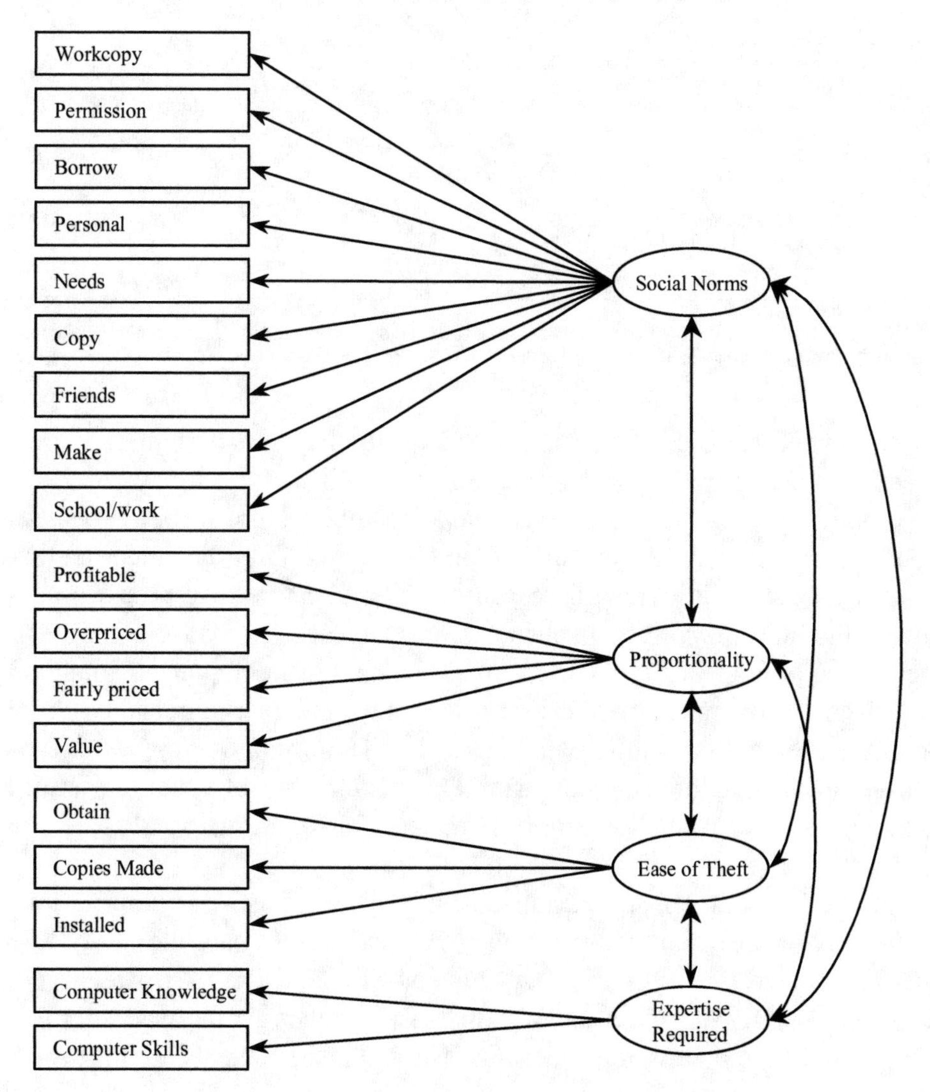

Table 2: Comparison of relative fit statistics for measurement and structural models of software piracy

Measurement Models

	Model Description	Chi-square Statistic	D.F.	Differences In Fit	Differences In D.F.	X2/df	GFI
1.	18 indicators, 4 latent factors, no error correlations.	369.20	129	----	----	2.862	.922
2.	12 error correlations allowed between indicators tapping similar perceptions.	157.11	117	212.09**	12	1.343	.968

Structural Models

	Model Description	Chi-square Statistic	D.F.	Differences In Fit	Differences In D.F.	X2/df	GFI	
3.	Simultaneous est. of meas. and struc. models. Direct effects from exogenous measures.	464.13	313	----	----	1.483	.950	.288
4.	Introduce direct effects from measurement structure.	464.13	313	no change				.411
5.	Delete direct effects from several demographic measures.	473.04	323	8.91	10	1.465	.949	.402
6.	Delete direct effects from proportionality and ease of theft factors.	474.83	325	1.79	2	1.461	.949	.413
7.	Add direct effects from proportionality to social norms and easy to knowledge.	480.24	327	5.41	2	1.468	.948	.413

Probability level: *<.05; **<.001.
Note: Models 5, 6, and 7 are significantly improved due to the additional degrees of freedom saved by altering the direct effects allowed in the model structure. GFI is the goodness of fit index provided by LISREL.

analyses since they did not sufficiently cohere to their assumed latent structures, social norms and expertise, or the remaining factors in the model.

Figure 1 depicts the measurement model relating to software piracy perceptions. Proportionality, for instance, represents an unobserved construct that generates the structure of relationships among its indicators--overpriced, profitable, value, and fairly priced. Each indicator is a linear combination of the latent measure proportionality, plus a random measurement error component. The initial model estimated assumes that these measurement errors are uncorrelated with the latent unobserved construct or with one another. The program computes asymptotically unbiased and efficient maximum likelihood estimates of parameters, as well as a likelihood ratio statistic that approximates a chi-square distribution in large samples.

Table 2 presents chi-square statistics and other relative fit comparisons for a model of software perceptions and piracy. As expected, the baseline model does not fit the data well ($\times^2$=369.20; d.f.=129; ratio=2.862; GFI=.922). In relatively large samples a general rule-of-thumb is that the chi-square/d.f. ratio should be less than 2.0 and the goodness of fit index (GFI) should exceed 0.95. While the results

Table 3: Measurement model of software pirating perceptions (N=523; chi-square=157.11; d. f.=117; GFI=.968)

Indicator Variable	Proportionality		Ease of Theft		Social Norms		Expertise Required	
Overpriced	1.000 [0.953]	(0.000)						
Profitable	0.393 [0.397]	(0.049)						
Value	0.575 [0.552]	(0.058)						
Fairly Priced	0.746 [0.722]	(0.064)						
Obtained			0.662 [0.542]	(0.064)				
Installed			0.874 [0.719]	(.0730)				
Copies Made			1.000 [0.860]	(0.000)				
Needs			0.794 [0.709]	(0.043)				
Workcopy					0.722 [0.712]	(0.039)		
Copy					0.805 [0.730]	(0.042)		
Permission					0.717 [0.635]	(0.048)		
Borrow					0.753 [0.696]	(0.042)		
Friends					0.684 [0.557]	(0.052)		
Personal					1.000 [0.856]	(0.000)		
Make					0.940 [0.837]	(0.040)		
School/work					0.647 [0.556]	(0.049)		
Computer Knowledge							0.595 [0.500]	(.177)
Computer Skills							1.000 [0.809]	(0.000)
Reliability		0.7672		0.7409		0.9013		0.4328

Note: Maximum-likelihood coefficients reported first. Standard errors reported in (), standardized solution reported in brackets. GFI is a goodness of fit index provided by LISREL.

from Model 1 approach the GFI threshold, they do not approach the appropriate ratio. However, this is largely because error correlations were not allowed between the observed indicators in the model. A review of the 18 indicators suggests that many of them tap similar beliefs, perceptions, or ideas even though they are not perfect replications of one another. Therefore, in Model 2 twelve error correlations are included, although the basic four-factor structure depicted in Figure 1 is retained. As expected, the fit of Model 2 is a dramatic improvement over Model 1 (ratio=1.343; GFI=.968). While other structural configurations were examined,

Figure 2: Structural model of software piracy

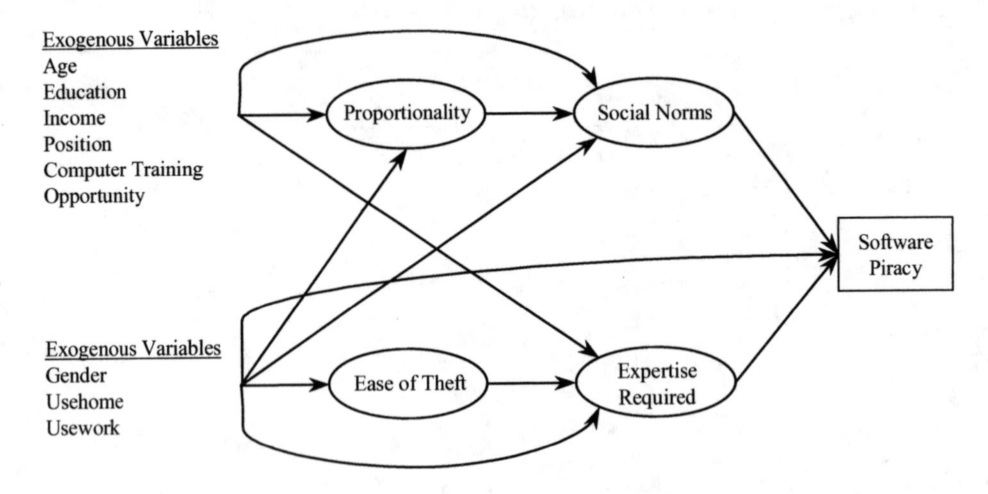

none were found to fit the data as well as Model 2. For instance, the latent factors of expertise and ease of theft might appear to overlap by looking at the indicators of each; however, the correlation between these structures is only .375. In addition, analyses combining the five indicators under one factor yield a reduction in the number of degrees of freedom but a large decrement in the chi-square statistic. Therefore, the four-factor structure is retained.

Empirical Assessment of the Structural Model

Due to its highly technical nature, a more detailed discussion of the empirical assessment of the structural model can be found in the appendix. This discussion tracks systematic changes to the proposed model and the resulting measures that assess goodness of fit. A summary description of this empirical assessment is provided in Table 2. Briefly, LISREL analysis was performed on 18 indicator variables comprising four latent factors. Various causal effects from both the latent factors and error terms were then systematically added or removed, and the resulting models, seven in all, were tested for relative goodness of fit. No improvement was possible from Model 6, which is displayed in Figure 2 and discussed next.

A four-factor structure was retained in Model 6. Four indicator variables (overpriced, profitable, value and fairly priced) loaded on the proportionality factor (see Table 1 for a more detailed description of the variables, and Table 3 for the maximum-likelihood coefficients and standard errors). The reliability measure was 0.7672. Four indicators (obtained, installed, copies made, needs) also loaded on ease of theft, with a reliability measure of 0.7409. The social norms factor was comprised of eight indicator variables (workcopy, copy, permission, borrow,

Table 4: Parameter estimates of structural model predicting software piracy (X2-480 .24, d. f.=327, GFI=.948)

Predetermined Variables	Endogenous Measures: Proportionality	Ease of Theft	Social Norms	Expertise Required	Software Piracy
Age	0.019	-0.024	-0.030	0.013	
	[0.154]*	[-0.025]	[-0.205]**	[0.179]*	
Sex	0.271	0.257	0.038	-0.082	0.310
	[0.100]*	[0.104]*	[0.012]	[-0.054]	[0.143]**
Education	0.091	0.053	0.079	-0.132	
	[0.064]	[0.041]	[0.048]	[-0.165]*	
Employed	0.015	0.029	-0.004	-.0013	
	[0.071]	[0.149]	[-0.017]	[-0.111]	
Income	0.038	0.096	-0.052	-0.045	
	[0.052]	[0.143]*	[-0.060]	[-0.108]	
University Position					
Admin.	0.334	1.260	-0.154	0.700	
	[0.063]	[-0.259]**	[-0.250]	[0.232]*	
Class. Staff	0.112	-0.487	-0.215	0.235	
	[-0.040]	[-0.192]	[-0.067]	[0.150]	
Prof. Staff	0.138	-0.471	0.142	0.327	
	[0.039]	[-0.147]*	[0.035]	[0.165]	
Faculty	0.001	-0.413	0.312	0.429	
	[0.000]	[-0.136]	[0.080]	[0.228]*	
Computer Experience					
Training	0.162	0.015	-0.092	-0.103	
	[0.123]*	[0.012]	[-0.060]	[-0.138]	
Home Use	0.044	0.003	0.019	-0.059	0.035
	[-0.155]*	[0.013]	[0.058]	[-0.366]**	[0.155]*
Work Use	.0002	0.040	-0.006	-0.035	0.034
	[-0.006]	[0.106]*	[-0.012]	[-0.152]	[0.103]*
Opportunity	0.094	0.217	0.248	0.034	
	[-0.096]	[0.240]**	[0.215]**	[0.061]	
Factors					
Proportionality			-0.316		
			[-0.271]**		
Ease of Theft				-0.263	
				[-0.426]**	
Social Norms					0.252
					[0.366]**
Expertise					-0.427
					[-0.301]**
R-squared	0.117	0.204	0.273	0.683	0.413

Note: Standardized coefficients in [ ]. Significant effects: *p <.05; **p < .001.

friends, personal, make and education) and yielded the highest reliability measure at 0.9013. Finally, two variables (computer knowledge and computer skills) loaded on expertise required. The reliability measure, 0.4328, is much lower than the three previous measures, but fairly typical of two-item measures.

Table 4 presents the maximum-likelihood and standardized coefficients for the model depicted in Figure 2. The final column of this table depicts the five direct effects on software piracy. Men are significantly more likely to pirate software than women as are those individuals who report using computers in their home or at work. In addition, those individuals who agree that using unauthorized software is not really theft are significantly more likely to do so than their contemporaries. Of these five direct effects, the standardized coefficients suggest that the strongest effects come from social norms and expertise. Therefore, the relatively high r-squared is largely the result of a person's perceived norms relating to computer software usage and expertise with computers.

Of the four factors in the model, only social norms (sc=0.366, $p<0.001$) and expertise required (sc=-0.301, $p<0.001$) were found to have significant direct effects on reported software piracy. The positive standard coefficient (sc) on social norms indicates that the more one views piracy as acceptable, the more one is likely to engage in this behavior. Similarly, the negative standard coefficient on expertise required suggests that the greater the perceived difficulty of making illegal copies, the less likely the behavior. Proportionality affected software piracy indirectly through social norms (sc=-0.271, $p<0.001$). This result suggests that the greater the perceived proportional value of software, the less likely one is to view piracy as acceptable. Ease of theft also affected software piracy indirectly. The significant direct effect on expertise required (sc=-0.426, $p<0.001$) indicates, as common sense would predict, that as the perceived ease of making illegal copies increases, less expertise is required to pirate software.

Turning next to the exogenous variables specified in the model, age was positively related to proportionality (sc=0.154, $p<0.05$) and expertise required (sc = 0.179, $p <0.05$) and negatively related to social norms (sc = -0.205, $p < 0.05$). This suggests that, as we pass our twenties, we are more likely to appreciate the proportional value of software and the expertise required to make illegal copies, and less likely to condone piracy behavior. We also found significant effects for gender, with males more likely to view piracy with greater proportional value (sc =0.100, $p<0.05$) and ease of theft (sc=0.104, $p<0.05$) and more likely to engage in self-reported piracy (sc = 0.143, $p < 0.001$).

Interestingly, computer experience worked in opposite directions. Formal training in computers was positively related to the perceived proportional value of software (sc = 0.123, $p < 0.05$), but using a computer at home was negatively related (sc = -0.155, $p < 0.05$). Increased home use of computers also lowered impressions of the expertise required to pirate software (sc=-0.366, $p<0.001$) and had a positive and direct effect on self-reported piracy (sc=0.155, $p<0.05$). The opportunity to pirate software was positively and highly significantly related to both the ease of theft factor (sc=0.240, $p<0.001$) and social norms (sc=0.215, $p<0.001$). This indicates that those with a greater opportunity to pirate software view the action as less difficult and more acceptable.

DISCUSSION

Microcomputer software is protected under US Code, Section 17, of the copyright law (Mason, 1990). Although the maximum penalties for copyright infringement have recently been increased, results from this and other studies confirm that a high proportion of people believe the behavior is permissible. Thus, empirical research or theory that begins by assuming software piracy is universally accepted as inappropriate behavior fails to recognize the attitudes and evaluations

Starting from this vantage point, our research makes several contributions to the study of software piracy, ethics, and technology. First, the results support and extend those of previous studies concerning piracy behavior. A sizeable proportion of the respondents reported incidents of piracy, which is consistent with previous research (O'Brien & Solomon, 1991; Sacco & Zureik, 1990). Our results also indicate that gender affects reported piracy behavior, with males more likely to pirate software than females. However, while age has been reported in some studies as having direct effects on piracy, our findings indicate that age is related to piracy behavior indirectly by significantly affecting three of the four endogenous factors making up the measurement model.

Second, by modifying the theories of reasoned action (and the Vallerand et al. extension applied to moral behavior) and planned behavior, the present investigation had a firm foundation on which to develop a model of software piracy. The factor representing social norms, for instance, is the strongest predictor of pirating behavior. Although survey questions addressed both attitudinal (personal) and normative (social) criteria, separate factors failed to emerge. This finding coincides with those studies (Miniard & Cohen, 1981; Ryan, 1982; Shephard & O'Keefe, 1984) that challenge the independence of the attitudinal and normative factors. Further, we also identify several antecedent variables that are likely to affect an individual's normative beliefs and, in turn, their social norms. In agreement with the theory of planned behavior, we find that perceived behavioral control (expertise required) has an important direct effect on self-reported piracy.

Third, certain conceptual distinctions characterizing software and other forms of intellectual property were investigated. One's perception of proportional value was found to be indirectly related to software piracy. Proportionality was negatively related to social norms; that is, if individuals perceived the price of software to be unfair they were more likely to report social norms in favor of software piracy. To our knowledge, this was the first attempt to investigate the concept of software as nonexclusive property. We operationalized the construct as the difference in an individual's attitude towards the theft of tangible versus intellectual property. The greater this perceived difference, the more likely the individual was to report social norms in favor of software piracy. Nonexclusivity, although found to be positively related to social norms, did not survive the model construction phase of our analysis. Thus, future investigations should develop a variety of means by which to investigate this concept since it would seem to have a common-sense relation to deviant behavior.

Understanding the attitudes and perceptions of those who pirate software may point toward those areas holding the greatest promise for solutions. For example, if people copy software because the price violates their sense of proportional value, software firms may consider two solutions: raise the perceived value of the product through marketing efforts or lower the price (Zeithaml, 1988). Similarly, if existing social norms are contributing to the growth in unauthorized software, several remedies are available to software manufacturers. Two such solutions include (1) changing the image offenders have of the industry through a public relations campaign or (2) encouraging institutional customers to develop software policies which discourage unauthorized copying. To our knowledge, software associations have not considered this first alternative. They are, however, actively engaged in the second. Software associations are hard at work, both in and out of court, to establish standards and guidelines for their institutional customers. Interestingly, our results suggest that part of this effort may be misguided. We find no relationship between awareness of employer's software policies and reported piracy. Similarly, Taylor and Shim (1993) also report no relationship. Reid et al. (1992) found no relationship between awareness of copyright law and unauthorized copying. However, before we can confidently exclude such policy considerations from the model, a more in-depth analysis across different institutions, particularly nonacademic institutions, is required.

Companies that develop commercial software applications must carefully consider the issue of copy protection. Protection methods, which range from dongles to key diskettes and access codes, complicate the product, add additional cost, and may require added support. However, when done correctly, the company may reap the rewards from increased sales of its product. Advice on copy protection methods generally touch on several points, including choosing a method that is difficult to "crack," simple to apply, requires a minimum amount of technical support, and does not involve special manufacturing techniques.

It's important to point out that this model is an individual-level one, which only tangentially addresses many of the important macro-level issues of software piracy. One of the primary issues concerning intellectual property involves whether legally protecting it encourages or stifles innovation. As one scholar argues, "The fundamental bargain made for either patent or copyright protection is disclosure to the public in return for a monopoly of either limited duration (for patents) or limited scope (for copyrights) (Davidson, 1989, p. 163)." The debate is far from resolved in favor of protection. Conner and Rumlet (1991) conclude that not protecting software may, paradoxically, increase profits while lowering the cost for the consumer. Pirates may actually create a market for a particular type of software by making it the operating standard in a particular organization or industry. While the piracy model reported here does not directly address the merits of legal protection, it certainly makes clear that consumers do not view software piracy the same as theft of tangible goods, regardless of whether such activity is illegal or against company policy.

A related debate concerns the ability of legal standards to actually protect intellectual property from unauthorized copying. For example, software piracy is much more prevalent in foreign countries with weak legal protection of intellectual property rights (Weisband & Goodman, 1992). An interesting area for future research would be a cross-cultural examination of attitudes and behavior regarding software piracy in countries with varying legal protection (Swinyard, Rinne, & Kau, 1990). Such research would be particularly illuminating if it were longitudinal. As many countries adopt stricter intellectual property laws in order to comply with international trade standards, we could measure how legislation affects social norms regarding intellectual property.

There are several concerns that must be addressed before generalizing these results to other populations or other types of intellectual property. First, piracy was self-reported. Though we promised anonymity, we still could not ensure that all respondents were truthful. Second, attitude and value scales are difficult to validate and may have limited reliability (Grosof & Sardy, 1985). Third, there may be non-respondent bias; respondents may have been less likely to copy software than non-respondents. These concerns are somewhat allayed by the fact that the frequency of reported software piracy was quite high. Furthermore, ethical and legal concerns make it unfeasible to observe participants actually copying software illegally during an experiment. Thus, in spite of their weaknesses, surveys such as this one play an important role in the study of sensitive topics such as software piracy. The model proposed here is not meant to be definitive. It needs to be refined and tested with other populations as well as other types of intellectual property. The current research is meant to be a starting point for further work in this important and developing area.

ENDNOTE

[1] We raise this link with deviance research since the behaviors investigated with such perspectives are similar to software piracy in value of the item stolen (criminologists often ask whether individuals have taken anything less than $50) and seriousness of the crime (seriousness remains a vague concept but is often used to rank order level of criminal activity).

REFERENCES

Abbott, A. F. (1990). Developing a framework for intellectual property protection to advance innovation. In Rushing, F. W. and Ganz Brown, C. (Eds.), *Intellectual Property Rights in Science, Technology and Economic Performance*, 311-339. Boulder, CO: Westview Press.

Ajzen, I. (1985). From intentions to actions: A theory of planned behavior. In Kuhl, J. and Beckmann. (Eds.), *Action-Control: From Cognition to Behavior*, 11-39. Heidelberg: Springer.

Akers, R. F. (1994). *Criminological Theories: Introduction and Evaluation*. CA: Roxbury.

Bologna, J. (1982). *Computer Crime: Wave of the Future*. San Francisco: Assets Protection.

Briggins, A. (1998). Soft on software piracy? Your loss. *Management Review*, June.

Cloward, R. and Ohlin, L. (1960). *Delinquency and Opportunity*. New York: Free Press.

Conner, K. R. and Rumlet, R. P. (1991). Software piracy: an analysis of protection strategies. *Management Science*, 37, 125-139.

Cooper-Dreyfuss, R. (1989). General overview of the intellectual property system. In Weil, V. and Snapper, J. W. (Eds.), *Owning Scientific and Technical Information*, 17-40. New Brunswick: Rutgers University Press.

Davidson, D. M. (1989). Reverse engineering software under copyright law: The IBM PC BIOS. In Weil, V. and Snapper, J. W. (Eds.), *Owning Scientific and Technical Information*, 147-168. New Brunswick: Rutgers University Press.

Fishbein, M. and Ajzen, I. (1975). *Belief, Attitude, Intention and Behavior: An Introduction to Theory and Research*. Reading, MA: Addison-Wesley.

Gopal, R. D. and Sanders, F. L. (1998). International software piracy: Analysis of key issues and impacts. *Information Systems Research*, December.

Grosof, M. S. and Sardy, H. (1985). *A Research Primer for the Social and Behavioral Sciences*. New York: Academic Press, Inc.

Hayduk, L. (1987). *Structural Equation Modeling with LISREL*. Baltimore, MD: John Hopkins Press.

Hettinger, E. C. (1989). Justifying intellectual property. *Philosophy & Public Affairs*, 18, 31-52.

Holsing, N. F. and Yen, D. C. (1999). Software asset management: Analysis, development, and implementation. *Information Resources Management Journal*, July-September.

Im, J. H. and Koen, C. (1990). Software piracy and responsibilities of educational institutions. *Information & Management*, 18, 189-194.

Jackson, W. (1999). *Yo, Ho, Ho, and A CD ROM!* New Zealand Management, December.

James, G. (2000). Organized crime and the software biz. *MC Technology Marketing Intelligence*. January.

Johnson, D. G. (1985). *Computer Ethics*. Englewood Cliffs, NJ: Prentice-Hall.

Joreskog, K. G. and Sorbom, D. (1989). *LISREL VII. User's Guide*. Chicago: National Educational Resources.

Long, J. S. (1983). *Confirmatory Factor Analysis*. Beverly Hills, CA: Sage.

Masland, M. (2000). Software Piracy A Booming Net Trade. *MSNBC News*. Retrieved on the World Wide Web: http://www.msnbc.com/news/177396.asp?cp1=1.

Mason, J. (1990). Software pirates in the boardroom. *Management Review*, 40-43.

Matsueda, R.L., Gartner, R., Piliavin, I. and Polakowski, M. (1992). The prestige of criminal and conventional occupations: A subcultural model of criminal activity. *American Sociological Review*, 57, 752-771.

Matza, D. and Sykes, G. M. (1961). Juvenile delinquency and subterranean values. *American Sociological Review*. 26, 712-719.

Miniard, P. W. and Cohen, J. B. (1981). An examination of the Fishbein-Azjen behavioral-intention model's concepts and measures. *Journal of Experimental Social Psychology*, 17, 303-309.

Nelson, R. R. (1995). Why should managers be thinking about technology policy? *Strategic Management Journal*. 16, 581-588.

Nesselroade, J. R. (1983). Temporal selection and factori invariance in the study of development and change. *Life Span Development and Behavior*, 5, 60-89.

O'Brien, J. A. and Solomon, S. L. (1991). Demographic factors and attitudes toward software piracy. *Information Executive*, 12, 61-64.

Parker, D. (1976). *Crime by Computer*. New York: Charles Scriber's Sons.

Reid, R. A., Thompson, J. K. and Logsdon, J. M. (1992). Knowledge and attitudes of management students toward software piracy. *Journal of Computer Information Systems*, 33, 46-51.

Ryan, M. J. (1982). Behavioral intention formation: The interdependency of attitudinal and social influence variables. *Journal of Consumer Research*, 9, 263-278.

Sacco, V. F. and Zureik, E. (1990). Correlates of computer misuse: Data from a self-reporting sample. *Behaviour & Information Technology*, 9, 353-369.

Samuelson, P. (1989). Innovation and competition: Conflicts over intellectual property rights in new technologies. In Weil, V. and Snapper, J. W. (Eds.), *Owning Scientific and Technical Information*, 169-192. New Brunswick: Rutgers University Press.

Schuler, C. (1998). Make Software Piracy Meet its Doom. CeePrompt! *Computer Connection*. Retrieved on the World Wide Web: http://www/ceeprompt.com/articles/030998.html.

Shepard, G. J. and O'Keefe, D. J. (1984). Separability of attitudinal and normative influences on behavioral intentions in the Fishbein-Azjen model. *The Journal of Social Psychology*, 122, 287-288.

Sutherland, E. H. (1947). *Principles of Criminology*. 4th Ed. Philadelphia: J.B. Lippincott.

Swinyard, W. R., Rinne, H. and Kau, A. K. (1990). The morality of software piracy: A cross-cultural analysis. *Journal of Business Ethics*, 9, 655-664.

Taylor, G. S. and Shim, J. P. (1993). A comparative examination of attitudes toward software piracy among business professors and executives. *Human Relations*, 46, 419-433.

Vallerand, R. J., Pelletier, L. G., Cuerrier, P. D., J. P. and Mongeau, C. (1992). Ajzen and Fishbein's theory of reasoned action as applied to moral behavior: A confirmatory analysis. *Journal of Personality and Social Psychology*, 62(1), 98-109.

Weisband, S. P. and Goodman, S. E. (1992). News from the committee on public policy: International software piracy. *IEEE Computer*, November, 87-90.

Wells-Branscomb, A. (1990). Computer software: Protecting the crown jewels of the information economy. In Rushing, F. W. and Ganz Brown, C. (Eds.), *Intellectual Property Rights in Science, Technology and Economic Performance*, 47-60. Boulder CO: Westview Press.

Zeithaml, V. (1988). Consumer perceptions of price, quality, and value: A means-end model and synthesis of evidence. *Journal of Marketing*, 52, 2-22.

APPENDIX

Indicators of the Measurement Model

The multi-context character of the survey, even though it remains a cross-sectional design, is significant since it affords the opportunity to empirically distinguish whether some of these perceptions pertain only to work/personal settings or are more general in nature. In addition, Table 1 presents information on the distributions of these 18 measures. The measures of skewness and kurtosis do not appear to evidence any dramatic departure from normality (Hayduk, 1987). Nevertheless, methods of analysis that do not require that all assumptions of multivariate normality are have been conducted and are referred to below.

Statistical Methods

LISREL was used to estimate both the measurement and substantive structural models. In analyzing the measurement (factor) model we empirically examine: (1) which observed variables are affected by which latent factors; (2) which pairs of factors are correlated; (3) which observed variables are assumed to be measured with error; and (4) which error component of the observed variables is correlated (Long, 1983). The LISREL program also affords the ability to estimate a covariance structure model that simultaneously specifies a factor and a structural equation, or causal, model. Factor models explain the covariation among a set of observed variables in terms of a smaller number of common unobserved, or latent, structures (Long, 1983; Nesselroade, 1983).

After achieving the most parsimonious factor structure, the measurement model is simultaneously estimated within a more complex recursive structural model that incorporates variables according to their perceived temporal priority. Moreover, these initial structures can be empirically tested against alternative nested models that can be compared by several statistical measures derived from the LISREL program.

As a whole, the present data are well suited to the type of analysis outlined above. However, several limitations of the data do exist. Since there is little substantive theoretical research upon which the present data were based, we examined a variety of alternative structures, using exploratory factor modeling, but present the confirmatory analysis for the most substantively plausible models. Second, since the data were collected using cross-sectional survey methods the investigators impose the temporal priority outlined in the structural model explored. Nevertheless, the included measures provide an excellent opportunity to examine the possibility of describing software piracy by using unobserved latent models of perceptions that may mitigate the effect of demographic measures. Moreover, in examining the proper structural configuration we are careful to note that this is an area of research that provides little theoretical guidance; therefore, we have investigated a variety of alternatives not presented in the tables.

Finally, the present sample is sufficiently large (N=523) to capitalize on asymptotic properties of statistical estimators, such as maximum likelihood. However, since the distribution of software piracy was found to be positively skewed, additional analyses were conducted using unweighted and generalized least squares methods that do not require that all of the assumptions of multivariate normality be met. The results were extremely robust and appeared to be unaffected by the method of estimation; therefore, the maximum-likelihood estimates are presented. Additionally, past research suggests that the presence of skewed or kurtotic variables makes the chi-square distribution larger than expected so that the results are actually more conservative than necessary (Hayduk, 1987).

Empirical Assessment of the Measurement Model

Both the standardized and maximum-likelihood coefficients are presented in Table 3. As evident in the table, each of the coefficients is several times larger than its standard error. Substantively the factors represent significant perceptions held by the respondents regarding their views of the propriety, ease, and skill needed to pirate software. For instance, expertise required is a combination of statements suggesting that people perceive that those who might be able to pirate software need to have special computer knowledge and skills, while ease of theft represents an underlying dimension implying that anyone with access to a computer can easily obtain and install software that is received in an unauthorized fashion. As outlined above, each of these factors appears to be a separate dimension surrounding software piracy. In addition, proportionality represents a dimension that taps the relative fairness of the profits and costs related to computer software. In contrast, the social norms factor appears to coalesce indicators that minimize the negative connotation, or consequence, of pirating software. For instance, the indicator "borrow" suggests that the copying of software is more like borrowing than theft, and "needs" implies that it is alright to copy software for particular business or professional needs. Thus, we appear to have a mixture of factors assessing the relative expertise of persons who might pirate software as well as the norms surrounding access to unauthorized software, whether this relates to the perceived fairness of software pricing or general norms of conduct.

Empirical Assessment of the Structural Model Predicting Piracy

The following analysis describes the outcome of the simultaneous estimation of the measurement model as part of a more complex recursive structural model predicting software piracy. Figure 2 presents a model that includes as exogenous measures (those that are determined outside the model) several demographic, employment, and computer abilities of the respondents and five endogenous measures (those that are causally dependent on other endogenous variables as well as the exogenous measures) including the measurement model. Figure 2 represents the final version of the model discovered through empirical analysis. However, the structural model began as a fully saturated recursive model; that is, each exogenous measure directly affected each endogenous measure, including piracy, and all four factors in the measurement model had a direct effect on the dependent measure of software piracy.

Model 3 represents the initial estimation of the fully saturated model. As the model becomes more complex, in comparison to the measurement model, there is a slight decrement in the ratio (1.483) and GFI (.95); however, each of these values are within acceptable ranges. Additionally, Table 2 also includes a column presenting the explained variance (r-squared, or R2) of the equation being investigated. Model 3 allows each exogenous measure to have a direct effect on software piracy, however, only five measures were found to have a significant impact. As the age of the respondent rises, the likelihood that they pirated software significantly declined. In contrast, being a male and using a computer at home or at work significantly increased the likelihood of software piracy. Moreover, software piracy appears to depend upon one's perception of opportunities to pirate.

In the next equation the factors making up the measurement model were allowed to have direct effects on software piracy.[1] While the overall fit of the model is not altered, the explained variance (R2) is significantly increased to .411. The addition of direct effects from the factor structure significantly mitigates the effects from the exogenous measures previously described. In fact, both the age and opportunity measures are found to no longer differ significantly from zero; i.e., thus, having no direct impact. Further, the effects from sex, usehome, and usework are halved but still significant at the (.05) level.

Due to these results, the direct effect of 10 of the exogenous measures to software piracy have been eliminated in Model 5 of Table 2. There is a slight decrement in the chi-square statistic of this model, however, by constraining these effects to zero we have saved 10 degrees of freedom. Therefore, overall, Model 5 is a dramatic improvement over Model 4. The only two factors that were found to have a significant and direct effect on software piracy were expertise and social norms; that is, if one perceives that a person requires expertise to pirate software they are significantly less likely to do so themselves and if one generally agrees that making copies of software is borrowing and not theft they are significantly more likely to possess unauthorized copies of software.

In Model 6 we investigate these issues further by eliminating the direct effects of proportionality and ease of theft on software piracy. As in the earlier discussion there is only a small decrement in the chi-square statistic but we gain two additional degrees of freedom. It is also important to note that even though proportionality and ease of theft do not have direct effects on software piracy it appears that they are significantly related to social norms and expertise, respectively. Therefore, the final model investigated in Table 2 restructures the measurement model slightly into one that coincides with that depicted in Figure 2. In this model proportionality has a causal impact on social norms while ease of theft has a causal relationship to expertise. The fit of Model 7 is not dramatically different from Model 6; therefore, by the rules of parsimony since Model 7 eliminates insignificant paths to software pirating it is preferred over the others investigated.

Table 4 also presents information regarding the effects of demographic, computer experiences, and university position measures on the four factors previously discussed. For instance age positively affects social norms but negatively impacts perceptions of expertise. Moreover, age is the only measure that significantly affects all four of the factors in the analysis. The only other measure that comes close to this impact is opportunity, which significantly affects three of the four factors.

The R-squared for each of the equations shows a wide variation in explanatory ability of the model. With regard to the reconfigured factor structure we can see that as proportionality, or the belief that software is fairly priced and accessible, increases, social norms corresponding to unauthorized use of software significantly decreases. Additionally, if one perceives that software theft is easy to do their belief that special computer skills are required to commit the theft is decreased. These are fairly straightforward findings that coincide with expectations, however, the particular temporal sequencing of the factor structure was unexpected.

ENDNOTE

1 This process is similar to a path model or stepwise regression that allows direct effects to be incrementally added or deleted from the model.

Chapter 10

Ethical Issues in Software Engineering Revisited

Ali Salehnia
South Dakota State University

Hassan Pournaghshband
Southern Polytechnic State University

The process of software development is usually described in terms of a progression from the project planning to the final code, passing through intermediate stages such as requirement analysis, system design, coding, system testing, and maintenance. One important aspect of these requirements concerns the reliability of the software. The use of computers for life-critical systems demands extremely high reliability of the computing functions as a whole. The consequences of negative results from unreliable systems and software are becoming public knowledge every day. Since these situations create a negative image for computer professionals and since these episodes create an environment of nontrust for the discipline, a good look at the ethical issues in software engineering is necessary. In this chapter, we look at each of the software engineering steps and the important aspect of their reliability and safety in the analysis, design, and implementation of software. We also examine the ethical aspects of the software and system development.

INTRODUCTION

As software become more complex and sophisticated, so too must the methods of writing these programs. Failure to take responsibility for errors will only mean more catastrophes. The price for these failures is rising even today and it will

Previously Published in *Managing Information Technology in a Global Economy*, edited by Mehdi Khosrow-Pour, Copyright © 2001, Idea Group Publishing.

be paid in the future by the computer professionals' through loose of dignity and trust (Lee, 1992). What follows are some examples of problems raised when ethical issues in software engineering were ignored.

In a local newspaper on May 20, 1996, we found an article with the following headline: "Bank error produces 800 near-billionaires." The story was about a programming error that increased the account balance by $924.8 million dollars for each of the bank's 800 customers. This is a total of $763.9 billion, which are more than six times the bank's assets.

On May 7, 1996 in another local newspaper we found an article with the title: "Error causes 2 jets to occupy same runway." This story explains how two passenger jets came within 1,500 feet of each other on the same runway because both were assigned the same flight number.

Another article titled, " Planes in Northwest lose link with air traffic control center," appeared on January 7, 1996 and the story explains how a regional center (part of a $1.4 billion computerized system) lost communications with an aircraft for a few seconds because of a software problem.

The following quotes were taken from a mug acquired at a 1982 ACM Computer Conference (Art101, 1982) in order to indicate to the reader how far we have come with software engineering steps and processes:

- Weinberger Law: "If builders built buildings the way programmers write programs then first woodpecker that came along would destroy civilization."
- Troutman's Programming Laws: "If a test installation functions perfectly all subsequent systems will malfunction; not until a program has been on production for at least six months will the most harmful error then be discovered; any program will expand to fill any available memory."
- Gioub's Laws of Computerdome: "The effort required to correct the error increases geometrically with time."
- Hare's Law of Large Programs: "Inside every large program is a small program struggling to get out."

The question is: What has changed or improved since 1982? Can we say that "the more thing change the more they stay the same?"

We believe that some of the problems in software development can be dealt with by computing professional if they are trained to explicitly practice ethical guidelines and accept their social responsibilities. Software developers are held responsible for the outcome of their software. Hence, they should also be held responsible if their design is at fault. Furthermore, they should assume total legal liability for their faulty and unreliable programs and they should be required to let their clients know when their systems fail to deliver something that is required of them, such as a missing function when the clients need it. Basili and Musa (1991) consider such an event as a reliability problem and therefore a failure.

ETHICAL ISSUES AND RESPONSIBILITIES

Each of the information and computing professional organizations including the ACM, DPMA, ICCP, IEEE, etc. has a code of ethics, which emphasizes responsibilities of software engineers. In the general Moral Imperatives section 1.1 of the ACM Code of Ethics we read. "As an ACM member I will ... contribute to society and human well-being.... an essential aim of computing professional is to minimize negative consequences of computing systems, including threats to health and safety. When designing or implementing systems, computing professionals must attempt to ensure that the product of their efforts will be used in society responsible ways, will meet social needs, and will avoid harmful effects to health and welfare" (1993).

We believe that ethical issues play a big role in the analysis and development of software products. Wood and et al. (1996) argue about the need for the information systems person to receive training in ethical implications and they indicate that the existence of professional codes of practices is a clear indication that ethical neutrality is not possible. They continue to argue "Self-reflection by systems analysis on the ethical implications of their practice should ensure that ethical decisions are not made implicitly for them."

A section of the IEEE Code of Ethics is stated as follows "We the members of the IEEE do hereby commit ourselves to the highest ethical and professional conduct and agree: To accept responsibility in making engineering decision consistence with the safety, health and welfare of the public, and, to disclose promptly factors that might endanger the public or the environment. " Since software developers are considered engineers and scientists, they should definitely abide by these guidelines and produce reliable and safe products. Explaining the importance of the standards in software engineering, Lee (1992) indicates, "The time has come to make software engineering a science rather than an art. Software standards must be codified and programmers must strictly adhere to those standards." Adding, "writing software programs is no less important than building a bridge, and it should be treated as such."

Balzer and Goldman (1979) and Pressman (1987) propose eight principles for good specification: Principle #1) Separate functionality from implementation; Principle #2) A process-oriented systems specification language is required; Principle #'3) A specification must encompass the system of which the software is a component; Principle #4) A specification must encompass the environment in which the system operates; Principle #5) A system specification must be a cognitive model; Principle #6) A specification must be operational; Principle #7) The system specification must be tolerant of incompleteness and augmentable; Principle#8) A specification must be localized and loosely coupled.

One problem with the traditional software development method is the specification changes during the system development. Such changes have implications that may affect all parts of the software, making the previous design inadequate (Johnson and Nissenbaum, 1995). A good design requires taking into account the business, personal, and social expectations of clients. Knowing the consequences of a faulty program can increase the communication and cooperation between the software design team and the users. This in turn can increase the reliability and quality of the software.

While technical expertise and know how are important in software design and development, understanding of the social systems in which the software is to be used is also very important. A majority of the computer science and information systems departments teach programming and software engineering courses only from the technical point of view rather than considering- the technical, organizational, sociological and ethical points of view. Huff and Finholt (1994) argue, "A dedication to a reliable product, a commitment to open dealing with clients, and a concern for including customer and employees in the design process are pans of both the ACM ethics code and of quality design."

SOFTWARE QUALITY AND RELIABILITY ACHIEVEMENTS

Software reliability can be measured or estimated by using historical and developmental data (Pressman, 1987). Software reliability is defined as "the probability of failure free operation of a computer program in a specified environment for a specified time" (Musa, Iannino and Okumoto, 1987). A software reliability model can be used to characterize and predict behavior important to managers and software engineers. While software failure can be defined as nonconformance to software requirements, Pressman (1987) believes that "all software failure can be traced to design or implementation problems. Further, he argues that "Reliability is the most costly, performance characteristic to assess and the most difficult to guarantee." It is very important to understand that the reliability of a computer program is an important element of its overall quality. If a program frequently fails to perform, it doesn't matters whether other software quality factors are acceptable or not.

Quality software is defined as: "Software that satisfies the user's explicit and implicit requirements, it is well documented, meets the operating standards of the organization, and runs efficiently on the hardware for which it was developed." Software quality may be divided into three measures: operability (accuracy, efficiency, reliability, integrity, security, timely fashion, and usability); maintainability (changeability, correctability, flexibility, and testability); and transferability (code reusability, interoperability, and portability). Ferdinand (1993) states that "One can correctly opine that many modern programming practices, along with the heavy data gathering that often accompanies them, are necessary to achieve quality products, but are in themselves not sufficient for reducing- defects or significantly improving the level of software productivity sought in software engineering."

One way to ensure software quality and to achieve reliability is to use formal methods, which means the user's needs must be expressed in a mathematical language. This technique is highly reliable. However, errors can be introduced into the design during the implementation process. Another way to insure the quality and the reliability of software is to institute Software Quality Assurance functions. Dunn and Ullman (1982) present a list of tasks that should be part of any software quality assurance plan: these tasks are: 1) System design; 2) Software requirements specification review; 3) Preliminary design review; 4) Detail design review: 5) Review of integration test plan; 6) code review: 7) Review of test procedure; 8) Audit of document standards: 9) Configuration control audit; 10) Test audit, 11) Define data collection, evaluation and analysis; 12) Tool certification; 13) Vendor and contractor oversight; and 14) Record keeping.

Also to ensure software quality and reliability the designers should consider and enforce software security. Describing computer security, Pfleeger (1989) indicates that the attack on software may occur by the act of modification, interruption, deletion, and interception. Changing a bit in its code can modify a program and may causes the system to crash. The three well-known categories of software modification can be listed as: 1) Trojan Horse—a program that overtly does one thing while covertly does another; 2) Trapdoor—a secret entry point to a program and 3) Program leaks—a program making information accessible to unintended people or programs (Pfleeger, 1989). Pfleeger (1989) also indicates that a program must be secure enough to exclude outside attack and must be developed and maintained so that one can be confident of the dependability of the program.

The ultimate goal of software quality is user satisfaction. Basili and Musa (1991) listing important attributes that satisfy users state, "The attributes most often named as significant are functionality, reliability, cost, and product availability date. Reliability often ranks first." However, some obscure errors can have disastrous consequences. These reviews are also mandated by the Department of Defense as part of the formal requirements for a contractor's quality assurance program (Johnson et al., 1995).

The third way to achieve high reliability is fault tolerance. Fault-tolerant computing is one method for increasing a system's useful lifetime. It should be not that for a given application, a system could be sufficiently reliable without fault tolerance. Furthermore, we would like to point out that using fault tolerant doesn't necessarily Guarantee that a system will be sufficiently reliable for a particular application. Availability is the probability that system will be operational at any given moment (Gantenbein, Shin and Wang, 1991). Software fault tolerance allows a system to detect errors and to avoid the failures, which result from them. If this objective is accomplished, errors cause little or no visible degradation of performance or reliability. The ultimate goal of software fault tolerance is to increase the reliability, availability, and or safety levels in critical applications.

Software fault tolerance (SFT) techniques fall into three groups: dynamic redundancy, fallback methods, and error isolation. The fault-tolerance techniques used in real-time control systems seldom conform to the basic paradigm of software fault tolerance, recovery block or N-version programming, but rather tend to be combinations of the three (Avizienis, 1985; Cha, 1986; Leveson, 1992).

Since the ultimate coal of software fault tolerance is to increase the reliability, availability, and safety of critical applications and since the ultimate goal of a CASE tool technology is to separate the software system design from the implementation of program code, the CASE tools should be very important for SFT systems development, while the classical approaches to software fault tolerant system does increase the complexity of the software systems.

Fault-tolerant strategies are concerned with keeping the software system functioning in the presence of errors. Strategies fall into three groups: dynamic redundancy, fallback method and error isolation (Nelson, 1989). One approach to dynamic redundancy is a processing technique known as voting. Data is processed in parallel by multiple identical devices and the output of these devices is compared. If a majority of the devices have the same result that result is assumed to be the correct one. Fallback or degraded-service methods are appropriate when it is essential for the software system to shutdown.

For instance, in a process-control system, if a software error is detected that is causing the system to fail; a separate back up piece of software may be loaded and executed to guarantee the safe shutdown of all processes being controlled by the system. The problem of common software faults given rise to similar errors in redundant software still remains open. With a careful analysis of a system design and requirement we can identify dangerous or difficult components of the software.

Requirement Analysis

A requirement analysis tool allows a system engineer to initiate, design, complete, modify, and maintain SFT systems. CASE tools can also provide special features for requirement analysis and documentation of the SFT systems. Functional requirement documents are often cast in a yen, specific, predefined format that identify each and even, requirement down to every minute detail. Each requirement in the requirements' specification list should have the features and constraints of each component along with the whole structure of the SFT systems, or/

and a technique for SFT systems such as RB, N-version Programming, CBS, or RNB.

Requirement trace ability is an important method of demonstrating SFT structures produced (or reliability of the product) to satisfy user requirements. Usually, this is demonstrated in steps by showing that the reliable SFT systems produced by the current development step can be traced back to the previous step of the software life cycle. For example, we should be able to trace back to the requirement specification of SFT structure from the design specification, or from the source code to the design specification.

The requirement of SFT systems trace ability at the code-back-to-design is shown by automatically creating a structure chart from the source and comparing this chart to the structure chart created during the design phase. Therefore, the requirement specification should show the structure of SFT systems, components of the SFT approach, and constraints of the SFT systems used in the project.

System Design Phase

Quality software is organized as a set of independent modules, each of which can be designed and tested separately. Each module views the other as black box with well-defined sets of inputs and outputs. Each module is accessed only through these inputs and outputs. This logical segregation of functionality is not only a key factor for ordinary software development but also a main factor of SFT systems implementation.

CONCLUSION

The process of software development is usually described in terms of a progression from the project planning to the final code, passing through intermediate stages such as requirements analysis, system design and coding system testing, and maintenance. One important aspect of these requirements concerns the reliability of the software. The use of computers for life-critical systems demands extremely high reliability of the computing functions as a whole. Applications that pace the state of art in this regard include spacecraft. Fly-by-wire systems for passenger aircraft, safety systems for nuclear reactor, and traffic control system for tracked vehicles and aircraft. The need for high reliability of software components of these life-critical systems has become more apparent with the increasing functionality being ascribed. One result of this increased functionality is the recommendation of various designs for achieving fault tolerant system.

The ultimate goals of a software system are to increase the quality, reliability, and availability, and safety of critical applications. Even though, the classical approaches to software quality, reliability, and fault tolerant systems increase the complexity and cost of large software systems, the implication of ethical issues can address same of these problems. An engineering approach with ethical issues in mind will enable the software developers to produce reliable software and in accordance with user requirements.

REFERENCES

ACM's Code of Ethics and Professional Conduct (1993). *Communication of the ACM,* 36 (12).

Art 101, Limited (1982). Atlanta, GA.

Avizienis. A. (1985). The N-version approach to Fault-Tolerance Software. *IEEE Trans. Software Eng,* 1 SE-11, pp. 1491-l497. Dec.

Balzer, R. and Goodman, N. (1979) . Principle of Good Software Specification, *Proc. of Specifications Software IEEE*, pp. 58-67.

Basili, V. and Musa, J. (1991). The Future Engineering of Software: A Management Perspective, *Computer, IEEE*, 24 (9), September.

Cha, S. (1986). A Recovery Block Model and its Analysis. *Proc. Fifth IFAC Workshop on o safety of Computer Control Systems (SAFECOMP 1986*), Oxford: Press, pp. 11-26.

Dunn, R, and Ullman, R. (1982). *Quality Assurance for Computer Software*, McGraw-Hill.

Ferdinand, A. (1993). *Systems Software and Quality Engineering*. Van Nostrand Reinhold.

Gantenbein, R., Shin, S. and Wang, Z. (1991). Software Fault Tolerance in a Distributed Real-Time Control System. *Proc. of 4th ISMM/IASTED International Conf. on Parallel and Distributed Computing and Systems*. Washin2ton D.C. Oct., pp. 61-64.

Huff, C. and Finholt, T. (1994). *Social Issues in Computing: Putting Computing in its Place*. McGraw-Hill.

Johnson, D. (1995). *Computer Ethics*. Prentice-Hall.

Johnson, D. and H. Nissenbaum (1995). *Computers, Ethics, and Social Values*.Prentice-Hall.

Lee, L. (1992). Computer Out of Control. *Byte*, Feb., p. 344.

Leveson, N. (1987). Building Safe Software, *Software Reliability: Achievement and Assessments* (B. Littlewood, ed). Oxford: Beackwell Scientific Publications, pp. 1-18.

Musa, J. D., Iannino, A., and Okumoto, K. (1987). *Engineering and Managing Software with Reliability Measures*, McGraw-Hill.

Nelson, S. (I989). Making the Business Case for CASE Technology, *Mainframe Update*.

Oz, F. (1994*). Ethics for the Information Age*. B&E Tech.

Pfleeger, C. (1989). *Security in Computing*. Prentice-Hall.

Pressman, R. (1987). *Software Engineering*, McGraw-Hill.

Shin, S. and Alishiri, Z. (1994). CASE Tools Comparisons. *Proceedings of the Association of Management*.

Shin, S., Salehnia, A. and Cong, B. (1993). Implementation of Software Fault Tolerant Systems, *Proceedings of 1993 International IRMA Conference*. pp. 466.

Summerville, I. (1989). *Software Engineering*. Reading, MA: Addison-Wesley.

Wood, A.T. et al. (1996). How We Profess: The Ethical Systems Analyst. *Communication of the ACM*, March, 39 (3), pp 69-77.

Chapter 11

Cyberspace Ethics and Information Warfare

Matthew Warren
Deakin University, Australia

William Hutchinson
Edith Cowan University, Australia

INTRODUCTION

We have seen a rise in computer misuse at a global level and also the development of new policies and strategies to describe organized computer security attacks against the information society–these strategies are described as being "information warfare." This is very different from the traditional view of attack against computers by the individual, determined hacker, a cyber warrior with a code of conduct to follow. Today the threats come from individuals, corporations, government agencies (domestic and foreign), organized crime and terrorists. This new world of conflict in the electronic ether of virtual cyberspace has brought with it a new set of ethical dilemmas.

COMPUTER HACKERS

In the beginning, there were hackers. A group of what seem now to be a simple case of technocentric juveniles out to challenge their wits against the system. The term "computer hacker" usually denotes those who try to gain entry into a computer or computer network by defeating the computers' access (and/or security) controls. Hackers are by no means a new threat and have routinely featured in news stories during the last two decades. Indeed, they have become the traditional

Previously Published in *Social Responsibility in the Information Age*, edited by Gurpreet Dhillon,

"target" of the media, with the standard approach being to present the image of either a "teenage whiz kid" or an insidious threat. In reality, it can be argued that there are different degrees of the problem. Some hackers are malicious, while others are merely naive and hence do not appreciate that their activities may be doing any real harm. Furthermore, when viewed as a general population, hackers may be seen to have numerous motivations for their actions (including financial gain, revenge, ideology or just plain mischief making). However, in many cases it can be argued that this is immaterial as, no matter what the reason, the end result is some form of adverse impact upon another party.

Steven Levy's book *Hackers: Heroes of the Computer Revolution* (1984) suggests that hackers operate by a code of ethics. This code defines main key areas:

- Hands-on imperative: Access to computers and hardware should be complete and total. It is asserted to be a categorical imperative to remove any barriers between people and the use and understanding of any technology, no matter how large, complex, dangerous, labyrinthine, proprietary, or powerful.
- "Information wants to be free." This can be interpreted in a number of ways. Free might mean without restrictions (freedom of movement = no censorship), without control (freedom of change/evolution = no ownership or authorship, no intellectual property), or without monetary value (no cost).
- Mistrust of authority. Promote decentralization. This element of the ethic shows its strong anarchistic, individualistic, and libertarian nature. Hackers have shown distrust toward large institutions, including, but not limited to, the state, corporations, and computer administrative bureaucracies.
- No bogus criteria: Hackers should be judged by their hacking, not by "bogus criteria" such as race, age, sex, or position.
- "You can create truth and beauty on a computer." Hacking is equated with artistry and creativity. Furthermore, this element of the ethos raises it to the level of philosophy.
- Computers can change your life for the better. In some ways, this last statement really is simply a corollary of the previous one. Since most of humanity desires things that are good, true, and/or beautiful.

During the 1980s and 1990s this pure vision of what hackers are was changed by the development of new groups with various aims and values. Mizrach (1997) states that the following individuals exist in cyberspace:

- Hackers (Crackers, system intruders)–These are people who attempt to penetrate security systems on remote computers. This is the new sense of the term, whereas the old sense of the term simply referred to a person who was capable of creating hacks, or elegant, unusual, and unexpected uses of technology.
- Phreaks (phone phreakers, blue boxers)–These are people who attempt to use technology to explore and/or control the telephone system.

- Virus writers (also, creators of Trojans, worms, logic bombs)–These are people who write code which attempts to a) reproduce itself on other systems without authorization and b) often has a side effect, whether that be to display a message, play a prank, or destroy a hard drive.
- Pirates–Originally, this involved breaking copy protection on software. This activity was called "cracking." Nowadays, few software vendors use copy protection, but there are still various minor measures used to prevent the unauthorized duplication of software. Pirates devote themselves to thwarting these and sharing commercial software freely.
- Cypherpunks (cryptoanarchists)–Cypherpunks freely distribute the tools and methods for making use of strong encryption, which is basically unbreakable except by massive supercomputers. Because American intelligence and law enforcement agencies, such as the NSA and FBI, cannot break strong encryption, programs that employ it are classified as munitions. Thus, distribution of algorithms that make use of it is a felony.
- Anarchists–These are people committed to distributing illegal (or at least morally suspect) information, including, but not limited to, data on bomb making, lockpicking, pornography, drug manufacturing, and radio, cable and satellite TV piracy.
- Cyberpunks–Cyberpunks usually have some combination of the above, plus interest in technological self-modification, science fiction, hardware hacking and "street tech."

Mizarch (1997) determined that new groupings with cyberspace had altered the initial code of ethics, which in the late 1990s was more concerned with:

- "Above all else, do no harm." *Do not damage computers or data if at all possible.*
- Protect privacy. *People have a right to privacy, which means control over their own personal (or even familial) information .*
- "Waste not, want not." *Computer resources should not lie idle and wasted. It's ethically wrong to keep people out of systems when they could be using them during idle time.*
- Exceed limitations. *Hacking is about the continual transcendence of problem limitations.*
- The communication imperative. *People have the right to communicate and associate with their peers freely.*
- Leave no traces. *Do not leave a trail or trace of your presence; don't call attention to yourself or your exploits.*
- Share! *Information increases in value by sharing it with the maximum number of people; don't hoard, don't hide.*

- Self-Defense *against a cyberpunk future. Hacking and viruses are necessary to protect people from a possible Orwellian "1984" future.*
- Hacking helps security. *This could be called the "Tiger team ethic": It is useful and courteous to find security holes, and then tell people how to fix them.*
- Trust, but test! *You must constantly test the integrity of systems and find ways to improve them.*

This newer code of ethics is more based upon the view that hackers are helping in the development of the information society and adding to its distinct nature. The ethics of imposing these values on others who are unwilling "victims" does not seem to be questioned.

INFORMATION WARFARE

The advent of the contemporary concept of "information warfare" (see Schwartau, 1994, 2000; Denning, 1999; Waltz, 1998) has raised tampering with computer systems to a new dimension. The individualistic, anarchistic, and rather naive actions of young hackers have been replaced by the determined, methodical, and organized workings of states, corporations, and criminal gangs. Initially, the term "information warfare" was concerned with damaging a country's national information infrastructure (NII) (Schwartau, 1994). For the purposes of this paper, the NII is defined as the physical and virtual backbone of an information society and includes, at a minimum, all of the following (Cobb, 1998):

- government networks–executive and agencies
- banking and financial networks–stock exchanges, electronic money transfers
- public utility networks–telecommunication systems, energy and water supply (military and civil), hospitals, air traffic control and guidance systems, such as the global positioning satellite system and the instrument landing system, both common to commercial aviation
- emergency services networks (including medical, police, fire, and rescue)
- mass media dissemination systems-satellite, TV, radio, and Internet
- private corporate and institutional networks
- educational and research networks.

Information warfare is concerned with the full spectrum of offensive and defensive operations such as electronic warfare, cyber-terrorism, psychological operations and so on (Main, 2000). Mainly it has been associated with the so-called "revolution in military affairs" (RMA), although in the civilian world the increased reliance on information networks and, more specifically, has opened this type of "warfare" to anyone with the motivation to practice it. In some respects information warfare is a subset of the RMA, which is also concerned with the military application of new technologies to the "battlespace" (Cobb, 1998), such as stealth, precision-guided munitions, and advanced surveillance capabilities.

Because of this military aspect of information warfare, its ethics have been associated with those of national conflict. It is a development of the nature of warfare. The development of "total war", arguably started in the actions of Napoleon but certainly present in the Second World War, has been extended by the advent of information warfare. The distinction between military and civilian targets has been blurred. The Kosova conflict illustrated this. The bombing of a television station as a part of the Serbian military machine displayed the importance of the control of information in modern conflict (Ignatieff, 2000). At the state level, there is a tendency to attempt to develop "information superiority" over every competitor. International and national legal systems still have to catch up with this trend.

Increasingly the traditional attributes of the nation-state are blurring as a result of information technology. With information warfare, the state does not have a monopoly on dominant force nor can even the most powerful state reliably deter and defeat information warfare attacks. Increasingly non-state actors are attacking across geographic boundaries, eroding the concept of sovereignty based on physical geography.

There are also ethical implications between developed and less developed countries. In terms of information warfare, each society has its advantages and vulnerability. For instance, the USA has an enormous advantage in digital facilities from fibre optic communications to satellites and sophisticated software production. However, this advantage also adds vulnerability, as digital systems are susceptible to attack. The infrastructure of a "networked" society (e.g., power, water) is very exposed.

Developing countries at a lower level of development have an advantage of slower communication and processing systems, but this lack of sophistication lessens the vulnerability to an information attack.

The advantages of developed societies in the information "struggle" can be summarized as:

- Advanced infrastructure;
- Have the intellectual property rights to most advanced developments;
- Advanced technologies;
- Have control of large corporations;
- Advanced networked society possible, reliant on technology but infinitely flexible;
- Capable of information dominance strategy; and
- Dominate perception management industries, e.g., media.

The advantages of developing nations can be summarized as:

- Lack of vulnerable electronic infrastructure;

- Low entry costs to get into electronic systems development;
- Web based systems know no geographic boundaries (in theory), hence neither does "place" of company;
- Networked society based on "clans," difficult to penetrate; and
- Cheap labour, often with an educated elite, e.g., India.

The implications of interstate behaviours have yet to be fully worked out. Of course, it is questionable whether "ethics" per se are relevant in an environment where international law (which should reflect ethical issues) and national self-interest dominate.

Information warfare has also spilled into the corporate arena. Adams (1998) emphasizes the movement of state aggression from military to economic. However, this conflict can be said to have also moved to corporate to corporate conflict, and even from corporate to individual. The use of information warfare techniques can be seen as just another factor of business behaviour. Grace and Cohen (1998) illustrate the dilemma in business of behaving ethically. They argue that, all too often, arguments are polarized into two choices: behave unethically or fail. For many organizations, business is a form of competitive warfare. Hence, techniques applicable to the military and intelligence services (information warfare) are viewed as feasible choices, although Grace and Cohen assert that it is wrong to assume that competitors' tactics are designed to destroy rivals. However, they do argue for an "international legal and normative infrastructure ... The point is often lost in analogies with war, however, is that a great deal of this infrastructure already exists in private and public international law" (p.181). Whilst this may be the case for general business practices, the advent of cyber-space has created problems of definition and jurisdiction.

However, over the last few years, legislation in America, Europe, and Australia has allowed intelligence and law enforcement agencies to perform these operations. For instance, in 1999, the Australia Security and Intelligence Organization (ASIO) was allowed (with ministerial approval) to access data, and generally "hack" into systems (Lagan & Power, 1999). Therefore, legal hacking is the province of the authorities. Hardly an ethical public stance to take. In fact, it just emphasizes the efficacy of information warfare techniques to others.

The problem with the whole area of computer crime and the ambiguity of the law in this area is also confused by the potential international legal implications of foreign attacks. However, it is rare for these offences to be detected, prosecuted, or proven even within a state. Morth (1998) further argues that although this form of information warfare may be illegal in international law, it is only states that are covered by this, not individuals or companies (for which there is no international law in this area).

The perception of information warfare at the state or corporate level is to obtain as much benefit as possible and cause as much damage as possible to the "enemy." Thus the intent is usually destructive; this is very different from the hacker code of ethics. The reason for this is the emergence of cyberspace attack as the critical ingredient in a new witch's brew of strategic conflict capability. Information warfare tools and techniques allow the potential to destroy communications, information dissemination, and the functioning of critical equipment with no reference to geography or an ability to physically destroy anything in the conventional sense. Such a capability poses a whole new kind of threat to international stability (Molander & Siang, 1998). This destabilizing potential can equally have an organization or individual as its targeted victim.

It is interesting that in a survey of Australian IT managers carried out by the authors in 1999, 66% thought there was no threat from their competitors (Hutchinson & Warren, 1999). In other words, the thinking was that their competitors were ethical; in this area at least. However, the perception of competitor was probably a very narrow one. If the definition is to include "all those who wish to compete with your resource base or market share," then conservationists are competitors of mining companies and animal rights activists are competitors with fur traders. The picture is not so stable if these concepts are brought in.

Some of the comments received in the survey about the likelihood of an attack by competitors illustrate the point:

- "Government agency. No competitors."
- "Perhaps, but not yet possible."
- "Not competitors but malicious individuals."

It is interesting that a government department had the idea that they had no competitors. The impression given is that, at this moment in time, ethical business practice and/or the legal system will stop competitors from these aggressive acts. Of course, it could be lack of exposure to the techniques which has caused little consideration of the possibilities.

SOME SPECIFIC TECHNIQUES IN INFORMATION WARFARE

There are a number of ways information or information systems can be used to gain advantage over (or give a disadvantage to) another organization. Examples of some aggressive tactics are:

- Information can be manipulated or "created" (disinformation) to provide the target or its environment (for example, clients) a perception that develops behaviours detrimental to the target or beneficial to the attacker. At one level, this can be viewed as advertising, and, at another, deliberate deception.

- Information can be intercepted, thus giving the interceptor an advantageous insight into the target's strengths, weaknesses, and intentions.
- Information flows in the target organization can be disrupted or stopped, thereby interfering with the normal processes of the target, producing an advantage for the attacker, for instance, by bringing a server or network down.
- A target organization can be "flooded" with information, thereby slowing or stopping effective processing or analysis of the incoming information.
- Information can be made unavailable to a target organization by destroying the storage medium or cutting off the information source.
- Disrupting the availability of data or making the system produce incorrect/dubious output can lower the credibility of information systems.
- Confidential or sensitive information can be exposed to the public, clients, government agencies, and so on, thereby embarrassing or in other ways harming the organization.
- Physical attacks on IT or other components of the system can be made.
- Subversion of the people who operate the systems can be attempted.
- Physical destruction of information (erasure or overwrite) without harming the infrastructure components can be effected.
- Logic attacks (malicious code) on system components can be executed.

Obviously, many of these tactics are not pertinent to the contemporary business world (at least, not any ethically based corporate strategy) but they do give an idea of the range of possibilities open to an attacker.

FUTURE TRENDS

Information warfare has shifted the ethical issues from a naive group of technically oriented, young individuals (hackers, etc.) to the organizational arena. The implications of this shift from the relatively benign impact of individual behaviour to that of organized groups has raised the magnitude of the ethical issues surrounding them. They need to be considered at both the legal and policy-making levels of national, international, and corporate institutions before practices become entrenched or cause irreversible damage. However, the future is likely to be even more problematic. Some viewpoints consider that information warfare makes war more thinkable. It does not require that waging information warfare be either destructive or unjust. To the contrary, ethical notions of just war fighting will likely continue to provide a useful guide to behavior well into the information age (Khalilzad, 1999).

If the predictions of those such as Kurzwell (1999) are to be believed, the merging of people and machines will take on a new dimension. Although it can be said that the contemporary developed world is networked, the idea of a cyborg-world gives pause for thought. At the more mundane level, the advent of the wearable computer with its sensory enhancing peripherals makes real individuals the targets, rather than their computer systems. If these individuals with their infraread and hearing sensors, satellite positioning capabilities, face recognition and perceptive software (see Gershenfeld, 1999; Pentland, 2000) become truly networked, then the potential for damage from information warfare can be raised by significant orders of magnitude. Much as some people have become dependent on the mobile telephone, the awesome capabilities of personalized systems will make individuals dependent on them for interpreting every signal from the environment. The potential for perception management and confusion will make the ethical problems of today seem very minor.

To conclude we will look at the earliest research into ethics. Aristotle determined that it is by nature that some people are good, others it is by habit, and others it is by instruction (Aristotle, 1976). In this new information age humankind has developed new, incredible powers. Humankind has the ability to wage total war using the global information infrastructure there is no room left for "goodness."

REFERENCES

Adams, J. (1998). *The Next World War*. London: Hutchinson.

Aristotle. (1976). *The Ethics of Artistotle:The Nicomachean Ethics*. UK: Penguin Books.

Cobb.A. (1998). Australia's Vulnerability to Information attack: Towards a National information Policy, Strategic and Defense Studies Centre working paper No. 310. Australia: Australia National University.

Denning, D. E. (1999). Information Warfare and Security. New York: Addison-Wesley.

Gershenfeld. (1999). *When Things Start to Think*. Hodder and Stoughton, London.

Grace, D. and Cohen, S. (1998) *Business Ethics*–second edition. Melbourne: Oxford University Press.

Hutchinson, W. E. and Warren, M. J. (1999). Attacking the attackers: Attitudes of Australian IT managers to retaliation against hackers. *ACIS (Australasian Conference on Information Systems) 99*, December, Wellington, New Zealand.

Ignatieff, M. (2000). *Virtual War*. London: Chatto & Windus.

Khalilzad, Z, White, J and Andrew W. M. (1999). *MR-1015-AF–Strategic Appraisal: The Changing Role of Information in Warfare*. USA: RAND Publications.

Kurzwell, R (2000). The coming merging of mind and machine. In *Your Bionic Future: Scientific American*, 10(3), 56-61.

Lagan, B. and Power, B (1999). ASIO cleared to hack into computers. *Sydney Morning Herald*, March.

Levy, S. (1994). *Hackers: Heroes of the Computer Revolution.*

Main, B. (2000). Information wrfare: And its impact on the information technology industry in New Zealand. *Proceedings of the NACCQ*. Wellington, New Zealand.

Mizrach, S. (1997). *Is there a Hacker Ethic for 90s Hackers?* Retrieved on the World Wide Web: http://www.infowar.com

Molander, R. and Siang, S. (1998). The legitimization of strategic information warfare: Ethical consideration. *Professional Ethics Report*, 11(4).

Morth, T. A. (1998). Considering our position: Viewing information warfare as a use of force prohibited by Article 2(4) of the U.N.Charter. *Case Western Reserve Journal of International Law*, 30(2-3), 567-600.

Pentland, A. (2000). Perceptual intelligence. *Communications of the ACM*, 43(3), 40-44.

Schwartau, W. (1994). *Information Warfare: Chaos on the Electronic Super-highway*. New York, USA: Thunder's Mouth Press.

Schwartau, W. (2000). *Cybershock. Thunder's Mouth Press*. New York, USA.

Waltz, E. (1998). *Information Warfare–Principles and Operations*. Artech House, Norwood.

Chapter 12

A Conversation Regarding Ethics In Information Systems Educational Research

Mark Campbell Williams
Edith Cowan University, Australia

Should ethics be a significant importance in information systems educational research? In this chapter, I reflect on my heuristic and psychologically-oriented self-study concerning some ethical improprieties which I committed during the data collection phase of an information systems educational research programme. As part of this heuristic reflection, I engaged in a number of self dialogues in the form of a conversation between various characters. Reported in this paper is one of these dialogues, concerning broad issues of ethics and research and discussing the notion of wisdom, maturity, meaning, and virtue. I begin by asserting that this is even more so when considering research investigating and using new media, such as the world wide web, in which acceptable ethical practices have yet to be established and consolidated.

INTRODUCTION

From 1991 to 1995 I conducted qualitative research with the implicit hypothesis that *open discourse* could balance *technicism* in the University Business Computing classes taught by myself and three colleagues. I was progressively a tutor, then a lecturer, then the Course Coordinator facilitating the reform of a first-year Undergraduate Business Computing Course for the Bachelor of Business degree. The basic idea of the reform was to encourage communication and discourse about the meaning and purpose and wider societal implications of business

Previously Published in *Managing Information Technology in a Global Economy*, edited by Mehdi Khosrow-Pour, Copyright © 2001, Idea Group Publishing.

computing in addition to the narrowly technical aspects. I discontinued the research on realizing, through a reflective self-study, that I had acted unethically. To address this breach of ethics, I conducted an heuristic inquiry, from 1993 to 1996, to delve deeply, using heuristic reflection, into the nature, and possible healing, of the causes of my research short-comings both in theory and practice. The change in research approach and direction rested upon my growing awareness of the importance of ethics, reflective practice, art, symbolic interpretation of experience, and spirituality.

In this chapter, I present a conversation concerning the general issue of ethics in research, structured loosely around an idea of wisdom with some keywords of maturity, meaning and virtue. This discussion is presented in the form of a conversation between several persons. I coordinated an exchange of comments, concerning qualitative research, between Mr. Craig Standing, Dr. Peter Taylor, Dr. Peter Standen and myself. With their permission, I extensively edited, expanded, and added to the dialogue to change the themes and content to suit my own purposes. To protect my colleagues, and to clearly signal the fictive nature of the dialogue, I used the names Thales, Heraclitus, and Auguste Comte for the participants. Thales (around the early sixth century B.C.) was "one of the Seven Wise Men of ancient Greece" (Hammond, 1972, p. 179) and thus I use his name to present balanced moderation in the dialogue that follows this section. I use the name of Heraclitus (around 500 B.C.), a curious blend of systematic and also brilliantly intuitive thinker, named *the Obscure* (Gould, 1972), to present the more esoteric ideas in the dialogue. Because of Comte's (1798-1857) belief that objective scientific laws are the only basis for the social sciences, I have taken his name to present the positivist opinion that only explanations derived from objective examination of phenomena themselves should be considered worthwhile (Knapton, 1972).

A HEURISTIC CONVERSATION

Mark: Friends, let us begin our discussion of ethics when conducting educational research, especially when using web-based information systems.

Heraclitus: Why "especially when using web-based information systems?"

Mark: It seems to me that some folk, including those that are normally wise and ethical, act in an unusually "loose" manner when using the web. I have experienced cases when researchers, not only students, act with gay abandon in copying and plagiarising material from the web. It is my opinion that ethical procedures and practices take considerable time to become established and consolidated in new media and new situations. Also, some researchers are so concerned about research rigour that, it seems to me, they sometimes neglect important qualitative or intuitive human concerns. Rather than using the word "ethics," perhaps we should begin by considering the concept of *wisdom*

Thales: When using the word wisdom, I take it that you are talking from a value-laden framework that you think is preferable and can be used across multiple frameworks to guide the process of researching. I would like to hear more about this metaphor for I am thinking along the same line. Too many of us seem to be under a great burden, perhaps under the name of rigour, to achieve acceptability in academia.

Mark: In your first insight you are correct, and I agree with your second comment. I follow what is known as a virtue theory especially as it has been expounded by Alasdair MacIntyre (1984) in his book *Beyond Virtue: A Study in Moral Theory.* As I understand him, MacIntyre asserted that wisdom can only be lived out in a virtuous life, informed by multiple ways of knowing and learning, in community. As Bowers commented, community is: "based on some conception of how a good person acts" (1993, p. 87). Wisdom can be passed on only through a kind of apprenticeship in a way analogous to a bird teaching its young to fly. Wisdom, itself, seems to be part of the warp and weave of life bound up with constitutive forms of knowledge - rational and non-rational, conscious and unconscious, cerebral and somatic, logical and intuitive, emotional and holistic - these and more. Linguistically, it is primarily in the form of moral, metaphorical narratives which represent the community's stored understanding and wisdom. C. A. Bowers is much taken by these ideas, quoting MacIntyre's comment: "The narrative of any one life is part of an interlocking set of narratives" (cited in Bowers, 1993, p. 87). Bowers (1993) compared

Figure 1: Metaphor of Wisdom in Teaching, Learning, Researching and Writing

WISDOM IN TEACHING, LEARNING, RESEARCHING & WRITING

experience and knowledge together with the power of applying them critically or practically

MATURITY	*MEANING*	*VIRTUE*
being duly careful and adequate as a developed person - ESPECIALLY	pursuing that which is considered significant and important - ESPECIALLY	virtuous character or behaviour; a good moral quality, as justice - ESPECIALLY
1. BALANCE: a stability of mind; an equilibrium of instrumental action with communicative interaction	1. PURPOSE: intention, resolution, and a reason for endeavour	1. TRUTHFULNESS: esp. make known or reveal one's opinions, biases, interests, motivations and agendas (incl. unconscious)
2. THOROUGHNESS: not superficial; done with care, completeness and context	2. DIRECTION: a way or course for endeavour	2. HONESTY: truthful, fair and just in character and behaviour
3. RIGOUR: strictness and logical exactitude	3. INTEREST: a feeling of curiosity or concern, can be a passion	3. COMMITMENT: dedication and application in a pledge or undertaking

this understanding with the limited understandings of some computing experts. He asserted that these forms of knowledge (what I term wisdom): "cannot be made explicit and re-encoded to fit the digital technology required for computer-mediated communication" (p. 86, 87). When I use the term wisdom, I am thinking of the virtue of wisdom in this broad sense. Allow me to present pictorially some key words which may stimulate our discussion (see Figure 1).

Comte: I think *rigour* should have more promenance in the diagram. I would use rigour as a generic term (meaning strength) that applies to all aspects of educational research.

Mark: Welcome Auguste. Rigour is important, but I would still tend to agree with Thales' earlier insights that some researchers do seem to be working under an incredible, some would say unnatural, burden of rigour and that they may need to "lighten their load." Eisner (1992, p. 30) states that "in the context of investigation, scientific inquiry is an art. There are no algorithms to follow in either the conduct of research or the interpretation and writing of the results." In a similar vein, Calas and Smircich (1990), commenting on a book promoting reform in Management in higher education, asserted that the authors, while ostensibly lauding "difference and diversity," use a pervasive rhetoric of the words "rigour" and "quality" that actually exercise repression and impoverishment (1990, p. 704). However, in arguing for a lightening of the load of rigour, I am not suggesting carelessness.

Heraclitus: Sometimes passion for a research area can lead to carelessness. Even so, I contend that anyone who has anything relevant and interesting (hopefully) to contribute should be able do so - whether or not it is done sloppily or rigorously judged from a certain cultural standpoint. Research writing in the social sciences in France, for example, is considered sloppy if it is *not* rhetorically persuasive. For example, consider the work of Jacques Ellul, erstwhile Professor of Law and Institutions at Bordeaux University and a well-respected authority, notoriously careless in citing and justifying assertions but powerfully rhetorical. As Ellul (1990) asserts: "Analyses which appear very rigorous, which are built on statistics, and which make no reference to these [deeper] problems are the most dangerous. For they, too, are ideological, but they pretend to be purely scientific and have an appearance of strictness that one does not find in more rhetorical but more honest studies" (p. 36). It seems to me that, from a general English-speaking cultural viewpoint, research writing is considered careless if it *is* rhetorically persuasive. I am reminded of Habermas' warning about the "unavoidably rhetorical character of every kind of language, including philosophical language" (Dews, 1986, p. 161). The word sloppy is value laden but generally I think it is fair to say that it presumably means that which is unwise, or immature (i.e., unbalanced, with a lack of thoroughness or with little sense of purpose or direction or passion) or immoral (i.e., dishonest or uncommitted).

Mark: Well, I must confess that perhaps it was a form of passion that led me to become stressed, perhaps frenetic, during some qualitative research I conducted from 1991 to 1993. I became exhausted and thus careless, perhaps sloppy in the data collection phase, losing notes and audiotapes and not properly considering the effects on students of my introduction of innovative teaching-learning strategies. I decided eventually to discontinue the research.

Comte: It sounds as if you should have used a postitivist research approach, with careful experimental design, a consistent hypthesis, and systematic data collection. Ethics would have be taken care of by an ethics committee.

Mark: It would have made things simpler, Auguste. I had been using an interpretivist research approach based on unstructured qualitative data, from interviews, and I had followed emergent research questions. Due to an administrative mix-up, my research proposal was not presented to the ethics committee.

Comte: The lesson is clear: make sure any research is accepted by an ethics committee.

Heraclitus: Be careful here Mark. Auguste would allow you to attribute responsibility for your behaviour to others. An ethics committee exists merely to ensure that you build a framework into your research so that your may be truthful, just and fair, and do no harm to others. In fact, you can behave ethically without an ethics committee, and vice versa. The choice is yours.

Mark: I agree. There is more to ethics in research than just the ethics committee. I have had research proposals accepted by research committees - it seems to me an easy thing to get through. My breach of ethics was not something that an ethics committee could have picked up. My ethical failure was that I had *imposed* some teaching-learning innovations on students without first discussing the situation fully and gaining the student's *fully informed* consent. An ethics committee would have accepted my assurance of gaining student's informed consent, but I realised, on reflection, that I had not acted with sufficient ethical propriety.

Thales: Of course the researcher must always act with complete and utter honesty.

Mark: And to act with honesty, I would argue that researchers should become what Schon (1983) terms a *reflective practitioner*. I realised the importance of reflection only after two years of working on my research project. After much in-depth phenomenological reflection, I came to realise that I had been driven by a subtle, yet powerful, psychological urge to build up my ego by achieving a significant research result. During 1993, I realised that when conducting this research I had to endeavour, in good faith, to *live out* my concerns rather than merely follow the list of rules of the ethics committee. I had to take seriously the idea of including another way of knowing in listening to the voice of the unconscious. I had to be open and listen well to the voices of others in a communitarian sense. From 1993 to 1995, on becoming aware of the ethical deficiencies in the way I collected the data in 1992, I realised that I had to be very careful about how I used the material. Our previous discussion of MacIntyre's (1984) virtue theory forms a backdrop to what I think about ethical considerations.

Comte: It is not for science to teach maturity, meaning and virtue. It goes without saying that researchers must have enough maturity and morality to never influence the outcome of research. I don't know what you mean when you talk about the unconscious."

Mark: I refer to depth psychology when I mention the unconscious. In my own progress towards maturity and morality, I have been much helped by Jung's idea of individuation.

Comte: One must strive for objectivity. It is difficult for the social sciences to achieve the miracles of the physical sciences, but researchers must strive to use the methods of the physical sciences.

Heraclitus: What about that research in the social sciences which is difficult to replicate and, due to inherent subjectivity, almost impossible *not* to subtly influence the outcome? I am no positivist. In fact, I speculate that the whole idea of academic positivism became a device to lead the researcher to maturity and morality. In other words, to promote a type of wisdom. In the physical sciences, quantitative research was sufficiently limited and narrow that other researchers could replicate the experiment to test its validity. However, in the social sciences very few interventions could be reproduced in a way that could attest to the validity of the original research. Perhaps the intellectual community evolved elaborate rules to restrict the issue and help researchers resist the temptation to use immature rhetoric (or to lie for recognition, or whatever). The hope may have been that, if students were trained in academic (sometimes esoteric) rules, then they would develop their intellect and thus grow in wisdom (outworked in maturity and morality). However, we are all aware of the history of academic and scientific fraud. I have some anecdotal evidence in that, of the eight or so academics with whom I have spoken informally regarding their theses, three openly said that they committed academic fraud (and this was in three of the most respected universities in the Western world). There are other ways to lead people to wisdom - for example, the adoption of a moral framework which is often the outworking of a religious commitment.

Mark: Whatever the case, the post-Enlightenment thinkers who formulated the academic rules developed a rigid system of experimental quantitative rigour (Bowers, 1992). This restricted approach has been seriously questioned originally by neo-Calvinist philosophy in the early twentieth century (Doowerweerd,

1979) and then by postmodernism, poststructuralism, feminism (Caraca & Carrilho, 1994) and neo-Marxism (Bowers, 1992). I like the viewpoint of Caraca and Carrilho (1994) when they commended us to "think of knowledge and particularly scientific knowledge not as the application of a previously established method aimed also at achieving previously determined ends, but as an *invention* of strategies for discovery, legitimation and communication, for which the *rhetoric of science* may perhaps come to provide a rather unexpected picture for many" (p. 786). In my researching, I endeavour to explore ways to be rigorous and yet to be passionate. As Eisner (1992) recommended: "When you have a conviction about what you believe is important to study or how you think it should be studied, my advice to you is to pursue that conviction."

Comte: You have conveniently ignored the history of the last 500 years here! Scientific method, because it is an *open* system of knowledge acquisition, became popular precisely because it liberated us from moral philosophy that enslaved. Remember Galileo at the Spanish Inquisition? What about the role of religious "wisdom" in increasing the happiness of the human races - from the ritual sacrifices of the Aztecs through the Dark Ages to its peak in the Inquisition and the enslavery of the Third World in the 1500-1600s? Scientific method was one of the major ways we rid the world of these tyrannies, by enabling anyone to come up with openly debatable and verifiable answers to fundamental questions such as whether the earth is flat, or is the centre of the Universe. The realities espoused by wise folk such as witchdoctors, prophets and popes were found wrong. What other method could have achieved this?

Heraclitus: Auguste, I am surprised that you have not grasped the deeper currents of history. Critical theory, post-modernism and deep ecology have clearly revealed the deeply-embedded violence inherent in scientific consciousness. It is open only in so far as one treads a narrow epistemology. You should remember that it was the secular political inquisition which was the most barbaric (people often deliberately blasphemed at the crucial moment so that they would be tried by the more "reasonable" religious inquisition). And both these pale before the terrible inquisitions of the Enlightenment French revolution terrors through to the Nazi holocausts and right up to Pol Pot. The Enlightenment project, with science as its talisman, was originally empowering, but the seeds of its enslaving agenda were within its original mandate to dominate nature. Ghastly destruction has been wrought by scientistic positivism - not only in its application for the weapons of war. In one way, scientific and technological instrumental rationality has proven to be more "wrong" and more disastrous for planetary ecology than all the witchdoctors, prophets and popes put together!

Mark: I think that there is something important in what you say, Heraclitus. Adorno and Horkheimer (1990), Ellul (1964), Marcuse (1969), and a host of other major social philosophers from Nietzsche (1968) onwards drew attention to the way in which Enlightenment thinking is now a debased and enslaving hegemony of modernity. But Auguste is surely true in pointing us to the social and personal benefits of free critical thinking, in freeing generations of people from the enslaving hegemonies of the debased authoritarian religious and monarchist institutions of the eighteenth century right up to the present.

Thales: This may be true, but these historical and philosophical currents are rather beyond the scope of our discussion at the moment. I think that both Heraclitus and Auguste are arguing for a balance. Eisner (1992) reminds us that "scientific research writing, in the end, is a construction and the more artistic in character, the better." The debate is leading us to move from the narrow metaphor of rigour in research writing back to the fundamental questions of why we research at all. Of course, for a lot of academics and researchers, researching is a job and not much more - you do what you can get away with and what gives you promotion and acclaim.

Heraclitus: I also have found instances of this lazy approach, but other persons have something more like a *faith* in the scientific or Enlightenment project (Goudzwaard, 1979). They believe that this rational scientific method will progressively solve our social, cultural and even personal problems - all we need is more progress in scientific knowledge and concomitant technological wizardry! This results in a type of positivism which is close to a religious faith (Adorno & Horkheimer, 1992).

Mark: I prefer to place my faith elsewhere.

Thales: Hmmm . . . Perhaps we should return to the wisdom diagram (Fig. 1). Using this wisdom in teaching-learning-researching and writing metaphor, we should ask some key questions.

Mark: The central question becomes: "Does the teacher-learner-researcher-writer demonstrate *wisdom*?" The corollary questions become; "Does the teacher-learner-researcher-writer demonstrate *maturity* (i.e., balance and thoroughness and rigour) and *meaning* (a sense of *purpose* and *direction* and even, I would argue, a controlling and controlled *passion* - at least an interest - for the subject) and *virtue* (not only in not lying - *honesty* - but in "coming clean" in *self-disclosure* (Moustakas & Douglas, 1985, p. 50), revealing one's consciously understood world-view belief systems and interests in the subject as well as what can be gleaned of one's unconscious drives, emotions and motivations; and in having the *commitment* to continue). At certain times and in certain areas of researching it would be wise to subsume certain areas of maturity (rigour) for the greater purposes of meaning and morality - rigour is only part of maturity which is, in turn, only a part of wisdom. However, as one reviewer commented, someone might say that I use the previous sentence for politically-oriented strategic action to authorise careless research. In the end, I have come to the conclusion that the wise choice is the one that is consistent with what Guba and Lincoln (1989) term quality, or goodness of the research, or, for heuristic research what Moustakas (1990) terms validity.

Thales: Mark, in your research wisdom metaphor, you mention truthfulness, honesty, and commitment as elements of virtue. I take it that you are thinking of the ethical dimension of research. I concur with Habermas, restating Socrates before him, that we must discard the notion of objective knowledge, because all knowing is inextricably bound up in the life-world interests of the knower (Held, 1980, p. 297). I thus consider that if a researcher were to disclose his or her life-world interests, the reader could take this into account. A form of generalisable knowledge may thus emerge from this intersubjectivity. Nobel prize winner in economics, Gunnar Myrdal, recommends that the solution to the value-impregnated nature of social science is to state value judgements at the beginning, rather than to suppress them (Mahoney, 1993, p. 177). And I would add, the more the better.

Thales: Guba and Lincoln (1989, Chap. 4) spend a whole chapter on what they term the "twin failures of positivistic science," and the risks and redeeming features of constructivist research ethics.

Mark: I think that this is important. Guba and Lincoln (1989, chap. 4) point out that risks occur through: the intimacy of face-to-face contact; the difficulty of maintaining privacy and confidentiality; the possibilities of violation of trust; the need for open negotiations; and what to include/exclude in the framing of the case study. For example, in my study: I had to overcome a major personality clash with one of the other participant tutors; I had to be careful, prior to the exam, not to read any potentially critical material from student interviews conducted by the research assistant; I had to be careful to avoid violating trust by misinterpreting comments made by other tutors, or inadvertently passing on information to others in my University School; I had to engage in open discussions and be open to possibly negative feedback from participants; and I had to discern whether or not to include material from student learning journals, or from my own personal reflections and dreams.

Erickson (1986) notes that two basic principles are that participants in qualitative research need to be comprehensively briefed and informed about the research and carefully protected from risks. Pitman and Maxwell (1992) basically concur, stressing that ethical decisions are ongoing with one of the first decisions being whether the inquiry can actually be conducted. I have described how this question was crucial in the direction of my research, because I was ethically forced to the conclusion that I could not use most of my martialled research material to justify strong assertions about the influence or degree of technicism and open discourse in University Business Computing tutorials.

Pitman and Maxwell (1992) comment that for all major theorists in qualitative research, ethical considerations were imperative and even, for Guba and Lincoln (1990), action-oriented partnership is the motivating raison d'etre. In this spirit, I did endeavour to change the teaching-learning process in an on-going manner with key participants during the course of the research. Pitman and Maxwell also stress the obligations of the researcher both to the community studied and to the profession.

Thus I considered that my responsibility, to myself, to the participants and to the research community, was to tell my story, with as much maturity, meaning and virtue as I could. My hope was to thus bring out *some* results from the research. However, I did not consider it my responsibility to disclose *all* my major understandings or perceptions relevant to those results. I saw that one of my ethical responsibilities was to be caring and nurturing to all involved in the research, including myself. Thus, even if the events, actions and thoughts were relevant to the results that I was discussing, I was selective in what I chose to reveal. In this way, I not only safeguarded participants in the research (by anonymity and by selecting what to include), but also myself (Clandinin, 1993). I need to respect the proper boundaries for privacy of my self as well as other participants.

As knowing is inextricably bound up in the lifeworld interests, it is also inextricably bound up in the style of writing that one employs. Specifically, in autobiographical studies, ethics can be understood as nurture and care (Clandinin, 1993). Oakley (1992), in an autobiographical work, deliberately fictionalised certain areas to safeguard herself and those close to her. Clandinin (1993) states that this practice is of great importance in an interpretive style of research reporting. While I have not inserted fiction into research data findings, I chose to report only those events, actions and thoughts that I thought would not be too personally revealing.

Thales: Mark, do you think it not too personally revealing to disclose your relevant interests, biases, agendas, and worldview values and beliefs relevant to your research?

Mark: Not at all Thales, I welcome the question. Indeed, you may recall that I particularly mentioned this aspect in the truthfulness section of my research wisdom metaphor. Borrowing Tesch's (1990) phrase, my particular palette of research methodologies and methods sprung from my personal synthesis of what Husen (1994) asserts are the three major strands of the humanist paradigm, as distinct from the positivist paradigm, in education. The philosophical foundations for these three strands are, respectively, continental idealism, phenomenological philosophy, and critical philosophy (not least, the Frankfurt school, including Habermas). Continental idealism, influenced by German Idealism and Hegelianism, has a core concept that the humanities (including gestalt psychology) had their own logic of research oriented to *understanding*, whereas the natural sciences (including experimental psychology) has a logic of research orientated to *explaining*. Phenomenological philosophy, founded by Husserl, takes a holistic perspective to understand the entirety of a phenomenon, and to contextually and empathically understand human motives. I have previously described critical theory, at least its main variation of the Frankfurt School, in some depth - the emphasis is on understanding human beings in their socio-economic setting, at the same time having particular and unique motives.

During the whole of my research, continental idealism was a minor, but consistent and underlying influence on me, evident in my hints for the quest for a sacred science. In 1991 to 1993, during my initial qualitative research, I was influenced mostly by critical philosophy through my interest in Habermas' (1984, 1987) notions of technicism and discourse. From 1993 to 1996, on my discontinuing of the initial research, I resurrected the research data with an heuristic inquiry. I then used a form of psychological reflection based on Jungian dream interpretation to further explore an impression I gained from the heuristic inquiry. Using Walcot's (1993) typology of "a comparison of ways of knowing and inquiring," I began the research mostly critical approaches, and then moved towards interpretive approaches, while never having used positivistic approaches. That said, I find it difficult to exactly categorise my various approaches in the research.

Heraclitus: Mark, what of your epistemology and ontology?

Mark: That would need another long discussion. Suffice to say for now, that, using Hitchcock and Hughes (1989, p. 15) definitions, I am neither a realist nor a relativist in ontology, and neither an objectivist nor a subjectivist in epistemology. I am basically a flexible type of Berkeleian idealist who resorts to pragmatism where it seems to work (if that is not a tautology).

Comte: What use is knowing that when all we have to do is to be resolutely objective and impersonal, and to rigorously and systematically strive to discover the underlying laws of nature which, irrespective of personal opinion, are absolute?

Thales: It depends a lot on what is being investigated, and (which may be good for you to think about Auguste) on what are your motives.

Mark: Thank you friends, for joining in this discussion. I have been helped better to understand my own ethical practice. Perhaps others can be encouraged to tread a similar journey, especially if exploring new media such as the world wide web.

REFERENCES

Bowers, C. (1993). *Education, cultural myths and the ecological crisis: towards deep change.* Albany. New York: State University of New York.

Calas, M. B. & Smircich, L. (1990). Thrusting toward more of the same with the Porter-McKribbin report. *Academy of Management Review, 15*, 698-705.

Caraça, J. & Carrilho, M. (1994). A new paradigm in the organization of knowledge. *Futures, 26* (7), 781-786.

Clandinin, D. (1993, November). *Personal experience methods in research on teaching.* Paper presented at the 1993 International Conference on Interpretive Research in Science Education, Taiwan Normal University, Taipei.

Dews, P. (Ed.) (1986). *Habermas: Autonomy and solidarity*. London: Verso.

Doowerweed, H. (1979). *Roots of western culture.* Toronto: Wedge.

Douglas, B. G., & Moustakas, C. (1985). Heuristic inquiry: the internal search to know. *Journal of humanistic psychology 25* (3), 33-55.

Eisner, E. (1992). A slice of advice. *Educational Researcher,* June-July, 29-30.

Ellul, J. (1964). *The technological society.* New York: Knopf.

Ellul, J. (1990). *The technological bluff.* Michigan, U.S.A.: Eerdmans.

Erickson, F. (1986). Qualitative methods in research on teaching. In M. C. Wittrock (Ed.), *Handbook of research on teaching (3rd ed.),* (pp. 119-161). New York: MacMillan.

Goudzwaard, B. (1979). *Capitalism and progress.* Toronto: Wm. B. Eerdmans.

Gould, J. (1972). Heraclitus. In Field Enterprises Educational, The *World Book Encyclopedia,* (1972 Edition, Vol. 9, p. 185), Chicago: Author.

Guba, E. G. & Lincoln, Y. S. (1989). *Fourth Generation Evaluation.* Beverley Hills, USA: Sage.

Habermas, J. (1984). *The theory of communicative action: Vol. 1. Lifeworld and system critique of functionalist reason* (Thomas McCarthy, Trans.). Boston: Beacon.

Habermas, J. (1987). *The theory of communicative action: Vol. 2. Reason and the rationalization of society* (Thomas McCarthy, Trans.). Boston: Beacon.

Hitchcock, G. & Huges, D. (1989). *Research and the teacher.* London: Routledge.

Husen, T. (1984). *Research paradigms in education.* In T. Husen & T. N. Postiethwaite, (Eds.)*The international encyclopedia of education: Second edition,* (vol. 9, pp. 5051-5056.). Tarrytown, NY: Elsevier, Pergamon.

Knapton, E. (1972). Comte, Auguste. In Field Enterprises Educational, *The World Book Encyclopedia,* (1972 Edition, Vol. 4, p. 746), Chicago: Author.

MacIntyre, A. (1984). *After Virtue: A study in moral theory.* Notre Dame, IN: University of Notre Dame Press.

Marcuse, H. (1969) *Eros and civilisation.* London: Allen Lane & Penguin.

Moustakas, C. (1990). *Heuristic research: Design, methodology and applications.* Newbury Park: Sage.

Nietzsche, F. (1968). *The Will to Power* (Walter Kaufmann & R. J. Hollingdale, Trans.). London : Weidenfeld & Nicolson.

Oakley, A. (1982). *Taking it like a woman.* London: Flamingo.

Pitman, M. A. & Maxwell, J. A. (1992). Qualitative approaches to evaluation: models and methods. In M. D. LeCompte, W. L. Millroy & J. Preissle (Eds.),*The Handbook and Qualitative Research in Education,* pp. 729-770. San Diego: Academic Research Press.

Tesch, R. (1990). *Qualitative research: analysis types & software tools.* Hampshire, U.K.: Falmer Press.

Walcott, H. (1993). Posturing in qualitative research. In M .D. LeCompte, W. L. Millroy & J. Preissle (Eds.), *The handbook off qualitative research in education,* pp. 3-52. San Diego: Academic Press.

Chapter 13

Software Piracy: Are Robin Hood and Responsibility Denial at Work?

Susan J. Harrington
Georgia College & State University

INTRODUCTION

Despite the existence of laws and much publicity surrounding software piracy, it is widely believed that software piracy is commonplace (Eining & Christensen, 1991; Simpson, Banerjee, & Simpson, 1994). A recent study (i.e., Business Software Alliance, 1999) confirms that software piracy is increasing, with a 2.5 percent increase in piracy in 1998 over 1997, resulting in $3.2 billion in losses to organizations in the United States and $11 billion worldwide. Yet reasons why such illegal behavior continues to occur are lacking. While some attempts have been made at AACSB-accredited schools of business to incorporate ethics education into business programs, there is no knowledge of such education's relationship to actual behavior, nor is there knowledge on what exactly should be taught. Because previous educational, software-based safeguards, and attempts at raising awareness have failed to stop software piracy, some researchers (e.g., Simpson et al., 1994) believe that only when contributory factors are isolated can appropriate measures be taken to reduce software piracy. In addition, Watson and Pitt (1993) suggest that software piracy research lacks attention to individual factors, important for further understanding of the phenomenon.

Various accounts (see Figure 1) have cited reasons for computer abuse (i.e., the unethical use of computers) that includes software piracy. Thus this study, guided by existing ethical decision-making models, looks at these reasons for computer abuse behavior and relates these to individual characteristics in an at-

Previously Published in *Challenges of Information Technology Management in the 21st Century*, edited by Mehdi Khosrow-Pour,

Figure 1: Some reasons for computer abuse given by various sources

Purported Characteristics of Computer Abusers	Citation
Lacking in awareness of consequences	Baum, 1989; Ladd, 1989; Bloombecker, 1990b
Rationalizations for computer abuse	Krauss & MacGahan, 1979; Parker, 1989
Robin Hood Syndrome	U. S. Dept of Justice, 1989a, 1989b; Perrolle, 1987; Forester & Morrison, 1990
Economic gain	President's Council, 1986; Bloombecker, 1990a; Parker, 1983; Eining & Christensen,1991

tempt to understand the underlying causes of this persistent abuse. Specifically, this study looks at the individual factors of Responsibility Denial and "Robin Hood" syndrome.

ETHICAL DECISION-MAKING MODELS AND SOFTWARE PIRACY

Both generalized ethical decision-making[1] models and specialized software piracy models exist which contain components appropriate to the understanding of software piracy. Rest's (1986) and Jones' (1991) generalized models of ethical decision making form a foundation for the study of both situational and individual factors. Jones' (1991) model reviews the current ethical decision-making models and integrates them into one model, largely founded on Rest's (1986) model. This model suggests that ethical decision making is a four-component process: (1) recognize the ethical[2] issue, (2) make an ethical judgment or determine what is right or wrong, (3) establish ethical intentions, and (4) engage in ethical behavior. These components likely interact and do not necessarily occur in the order listed. Empirical support has been found for this model when applied to computer-related ethics issues, including software piracy (Eining & Christensen, 1991).

Ethical Judgment and Intent

Nisan (1984) suggests that ethical judgments consist of individuals' standards of behavior (their norms) and general principles regarding right and wrong. These general principles often rely on seriousness of consequences, number of others affected, etc. General ethics theories incorporate these principles and exist to explain the basis of peoples' ethical judgments. The exploration of ethical theories can be used to alter the quality of decisions being made regarding computer technology and has been readily incorporated into ethics education or training programs (Henry & Pierce, 1994).

However, Simpson et al. (1994) found no effect of individuals' ethical judgment/internalized norms concerning software piracy on the subjects' responses to whether they had ever pirated software. Similarly, Harrington (1995) in a study of IS employees found that, while 17 to 21 percent thought it was "OK to copy software" or not wrong to pirate software, 34 percent said they would copy software, a much higher percentage than said it was OK or not wrong. Such findings are significant, for if there is no effect of judgments on behavior, ethics education or training that focus on ethical theories and thus judgment may have no effect on unethical behavior (Simpson et al., 1994). Therefore it is important to confirm or refute these results and to understand further the individual characteristics that may contribute to a person's software piracy intentions that are contrary to the person's ethical judgments.

Vitell and Grove (1987) propose that a person may use neutralizations after an ethical judgment, making intentions or behavior inconsistent with judgment. Such neutralizations may be a rationalization for placing other values above ethical values. It is believed that the computer abuser often goes through a stage characterized by a decline in ethical judgments and a rationalization process that enables unethical intentions (Conger, Loch, and Helft, 1995). Therefore, it appears that ethical intent is susceptible to neutralizations and other influences. For example, personal wealth and security may be values placed above concerns for honesty and property rights. In effect, ethical intentions or behavior may differ from ethical judgment.

Responsibility Denial

As previously shown in Figure 1, rationalizations and lack of awareness of consequences have been cited as a source of computer abuse. A personality characteristic that describes such behavior has been previously discussed among social psychologists. Schwartz (1977) suggests that individuals differ in their awareness of consequences and may or may not feel personally responsible for others. He describes this personality characteristic as Responsibility Denial (RD). He suggests that RD, defined as the tendency to ascribe responsibility to oneself or to depersonalized others, is a relatively stable personality characteristic related to the acceptance or rejection of rationales for denying responsibility for the consequences of one's behavior. Those high in RD would agree with or offer rationalizations for denying responsibility. In other words, they would not accept responsibility for their actions. Staub (1978) suggests that those low in RD tend to accept responsibility and to be responsible for the welfare of others, live up to commitments, and follow either personal or societal rules and dictates. In support of this proposition,

Schwartz (1973) found that ethical judgment had no impact on altruistic behavior among those high in RD.

Kohlberg and Candee (1985) similarly suggest that "moral responsibility" is an individual characteristic that provides consistency between what one says one should or would do and what one does; it is a concern for and acceptance of the consequences of one's actions. Moral responsibility denotes follow through between one's ethical judgment and ethical behavior. They propose that responsibility is a second set of rules or criteria used by the individual to form an intention to "follow through." Therefore, we propose:

H1: Those high in Responsibility Denial (RD) will be more likely to pirate software.

Robin Hood Syndrome

Computer ethics literature (cf. Figure 1) also suggests that the Robin Hood syndrome may be related to higher levels of computer abuse. Robin Hood syndrome is the belief that harming a large organization to the benefit of an individual is the right behavior. Labeling of organizations as "the bad guy" may open the way for continuing hostility directed at the organizations (Snyder & Swann, 1978). Moreover, social science researchers (e.g., Staub, 1978; Kelman & Hamilton, 1989) suggest that dehumanizing or describing potential victims in negative terms disinhibits aggression toward the victims.

For those working within an organization, the "Robin Hood" perspective may be related to organizational commitment, which has been defined as the relative strength of an individual's identification with and involvement in a particular organization. Organizational commitment is believed to be a stable individual char-

Figure 2: Model under study

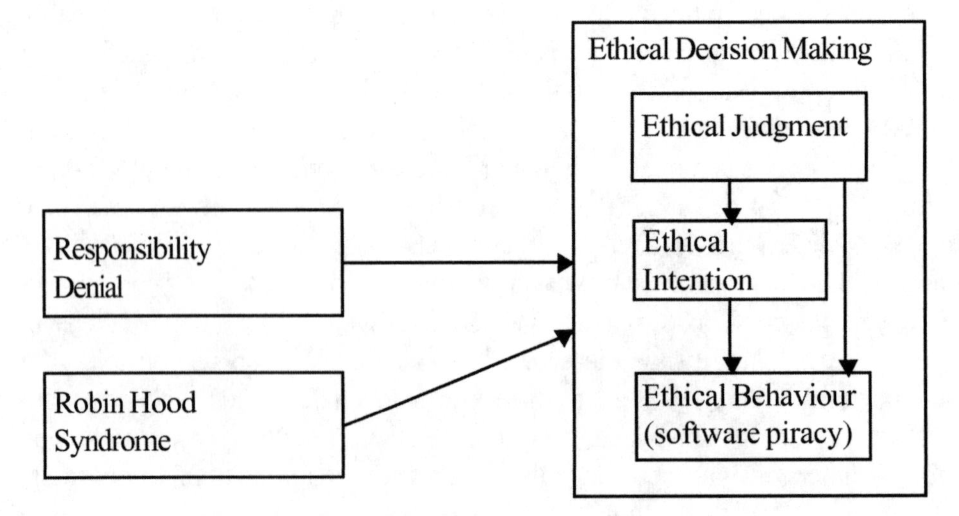

acteristic related to acceptance of the organization's goals and values, a willingness to exert effort for the organization, and a strong desire to be a member of the organization.

Therefore, it is a proposition of this study that Robin Hood may also allow individuals to neutralize ethical judgments about software piracy and copy software offered for sale by large organizations:

H2: Those high in Robin Hood Syndrome will be more likely to pirate software.

Thus this study will test the model shown in Figure 2 below.

METHOD

A questionnaire consisting of the measures for RD and Robin Hood, as well as a vignette describing software piracy, was administered to 102 information systems majors in a southern university. The age distribution was: 20 percent, 17-21 years old; 50 percent, 22 to 26 years; 12 percent, 27-31 years; 8 percent, 32-39 years; 8 percent, 40-47 years; and 1 percent, 48-57 years. The software piracy vignette (adapted from Harrington, 1989) described a friend who was copying an expensive software package and giving it away for free. Vignettes have the advantage of providing a less intimidating way to respond to sensitive issues and provide realistic scenarios that place the subject in a decision-making role. Moreover they avoid the subject's tendency to try to gain experimenter approval and so are commonly used in ethics research.

Measurement of Ethical Judgment, Intent, and Behavior

A vignette was used to measure the dependent variables of ethical judgment and ethical intent. Each vignette was followed by Likert-type questions asking how much the subject agrees or disagrees with statements describing the behavior portrayed in the vignette.

Disagreement to statements such as "Those who knowingly accept the illegally copied software packages have not done anything wrong" and the subject "would copy software for my use or for my friends" were used to measure ethical judgment and ethical intent, respectively. The subjects were also asked about their previous behavior with the questions, "Have you ever copied software, other than for backup purposes, that was copyrighted?" and "What percentage of your software (if any) is copied from a copyrighted source?" Hunt and Vitell (1986) suggest that weaknesses in existing ethical research occur when subjects are asked if they think persons in vignettes were ethical (or unethical) in their behavior, rather than being asked what they would do in the same situation. The statements used in this study avoid such problems and are consistent with the ethical decision-making models proposed.

Measurement of RD

The RD scale consisting of 28 items developed by Schwartz (1973) was used. Schwartz (1973) reports the RD scale of 28 items has an alpha coefficient of .78 to .81, a good reliability. Test-retest reliability of the original scale over a seven- to ten-month period and under different testing conditions was .81. Additional validity of the original instrument was supported by a correlation of -.01 with social desirability. Finally those low in RD behaved as expected in a game requiring cooperation and received significant and positively-correlated peer ratings on considerateness, reliability and helpfulness. Thus the RD scale has good validity.

After factor analyzing the responses to this scale, the current study found RD to consist of two major factors, which were labeled *Responsibility* and *Responsibility Denial*. The questions comprising these factors differed slightly in their perspectives. The first factor indicated whether people would take initiative when common practice may be to ignore the situation (e.g., "If a good friend of mine wanted to injure an enemy, it would be my duty to try to stop my friend."). The second factor indicated whether people would use excuses for unethical or insensitive behavior (e.g., "When you consider how hard it is for an honest person to get ahead, it is easier to forgive those who deceive others in business."; "I wouldn't feel that I had to do my part in a group project if everyone else was lazy.").

Measurement of Robin Hood Syndrome

There is no known measure of Robin Hood Syndrome. Therefore several Likert-style statements were constructed and factor-analyzed. The result was two factors, labeled *Robin Hood* and *Company over individual*. The *Robin Hood* factor consisted of the statements, "It is OK to take advantage of a big company whenever possible, even if it means harming the company" and "It is OK to take advantage of an individual whenever possible, even if it means harming the individual." While the responses to these statements seemed to vary in the same direction (i.e., Cronbach alpha was 0.79), the vehemence with which the subject responded often differed, often with stronger disagreement to the second question (involving the individual) over the first (involving the company). Therefore a difference score was computed by subtracting the two responses to see if the subject differed in the vehemence of the agreement or disagreement to these two statements.

The second factor, called *Company over individual,* consisted of the statements "In working for a company, I am willing to put in a great deal of effort beyond that normally expected in order to help the company be successful" and "Harming an individual is more wrong than harming a company" (reverse-scored).

Table 1: Comparison of Ethical Judgments, Intentions, and Behavior

	Agree (%)	Neutral (%)	Disagree (%)
Judgment: 1. "Those who knowingly accept illegally copied software packages have not done anything wrong."	6	6	88
Intention: 2. "I would copy software for my use or for my friends, too."	31	25	44
Behavior: 3. Have copied software, other than for backup purposes, which was copyrighted.	57		43

Behavior: 4. Percentage of your software that is illegally copied from a copyrighted source:	Percentage of Respondents
0 percent	37
Less than or equal to 1 percent	11
From 1 to 10 percent	35
From 11 to 50 percent	7
From 51 to 75 percent	4
From 76 to 100 percent	6

Table 2: Spearman Correlations

	1	2	3	4	5	6	7	8
1. Responsibility	(0.89)							
2. Responsibility	-0.19	(0.73)						
Denial	0.09							
3. Robin Hood	-0.13	0.13	-					
(as a difference)	0.23	0.24						
4. Robin Hood	-0.36	0.43	0.25	(0.79)				
(as a sum)	**0.001**	**0.000**	**0.02**					
5. Company	0.82	-0.12	-0.11	-0.33	(0.83)			
over individual	**0.000**	0.27	0.30	**0.002**				
6. Age	0.22	-0.12	-0.07	-0.24	0.26	-		
	0.05	0.30	0.50	**0.03**	**0.02**			
7. Ethical	0.36	-0.42	-0.16	-0.47	0.33	0.37	(0.81)	
Judgment	**0.001**	**0.000**	0.17	**0.000**	**0.003**	**0.001**		
8. Ethical	0.09	-0.12	-0.26	-0.20	0.16	0.29	0.66	(0.91)
Intent	0.42	0.28	**0.02**	0.08	0.16	**0.009**	**0.000**	
9. Percentage of	0.05	0.25	0.22	0.24	0.09	-0.27	-0.39	-0.56
software copied	0.66	**0.02**	**0.05**	**0.03**	0.43	**0.02**	**0.001**	**0.000**

First line is the correlation coefficient; second line is the p-value.
Boldfaced numbers represent significance levels less than 0.05
Cronbach alphas, where appropriate, appear in parentheses on the diagonal.

The Cronbach reliability of this measure is 0.83, and it appears to more closely measure the organizational commitment aspect of Robin Hood syndrome.

RESULTS

Table 1 presents the frequency of responses related to ethical judgment, intention and behavior. Of note is that more respondents had unethical intentions (31 percent) than those having unethical judgments (6 percent). This result is consistent with previous findings that people form intentions that differ from their judgments. Similarly, unethical behavior exceeded both judgment and intention with 57 percent having copied illegally.

Because those not owning a computer would not have the incentive to pirate software and would be less likely to empathize with the vignette, fifteen (15) individuals who did not own a computer were eliminated from further analysis. Table 2 shows the correlations between the variables under study using the 87 remaining individuals. The correlations between ethical judgment, intent and behavior are significant at the $p < 0.001$ level, showing a clear relationship between them consistent with the ethical decision-making models previously discussed.

Ethical judgment was significantly correlated with all independent variables proposed, except Robin Hood as a Difference. Ethical intent was significantly correlated with Robin Hood as a Difference and with Age. Unethical behavior was significantly correlated with Responsibility Denial, Robin Hood both as a Difference and as a Sum, and Age. A Mann Whitney U test, equivalent to a non-parametric T test (not shown) using the question "Have you ever copied software, other than for backup purposes, that was copyrighted?" showed that Responsibility, Responsibility Denial, and Company over Individual significantly differentiated software pirates from others at the $p < 0.05$ level.

Limitations

Although the study found a clear relationship between ethical judgment, intent, and behavior, the finding may be partially due to common method variance or same-source bias. Unfortunately, this potential bias is unavoidable. Nevertheless, the strength of the relationship between ethical judgment, intent, and behavior suggests that a relationship exists beyond that which may be caused by common method variance.

This study is also subject to the limitation of a student sample taken in one university in one region of the country. Thus the results may not be generalizable to IS students or personnel in other regions.

DISCUSSION AND CONCLUSIONS

This study hypothesized that Responsibility Denial and the Robin Hood syndrome influence IS students' unethical behavior regarding software piracy. The hypotheses were supported. Software piracy is clearly related to these characteristics. The characteristics also are related to unethical judgments suggesting that those high in RD or Robin Hood syndrome are more likely to see nothing wrong with software piracy. This finding is somewhat unexpected, for it is believed that RD and Robin Hood are more frequently used to rationalize behavior that is in opposition to judgments. It may be that those high in RD and Robin Hood syndrome may be denying that their behavior is wrong by rationalizing the unethical judgment itself. Alternately, those high in RD and Robin Hood syndrome may have low levels of moral development, a suggestion supported by the correlation of these variables with age. Further research is needed to confirm these possible findings.

This study also found that the ethical judgments of these IS students are generally ethical with respect to the software piracy. Only 6 percent said that there was nothing wrong with copying licensed software. This compares favorably to Harrington's (1995) study, which found that twenty one (21) percent of IS personnel did not believe software piracy to be wrong, as well as to a study by John Carroll (cf. Parker, 1983), who found that approximately 25 percent of students in 1977 believed it is ethical to use a program known to them to be proprietary in such a way as to avoid being charged for its use. While beyond the scope of this study, it is possible that the low percentage found in this study may be because these students have been exposed to business ethics courses and ethics modules in their IS programs. Therefore ethics education may be raising the awareness that software piracy is wrong.

However, it is important to point out how the findings of this study may help answer the question of whether ethics education can be improved. The fact that a large percentage of students know software piracy is wrong yet still have pirated software points to the fact that students will behave unethically even if taught ethical theory. The relationship found here between Responsibility Denial and software piracy points up the possibility that a focus on acceptance of responsibility and an awareness of the consequences of one's actions may reduce unethical behavior. The relationship between Robin Hood syndrome and software piracy also suggests that some do not understand that the organization consists of and is people and its ongoing viability will benefit those who work there. Therefore, interventions that encourage students to think about the good of society and the importance of organizations to society may also be a useful addition to some business ethics classes. While not all organizations are benevolent, it may be particularly helpful for those high in Robin Hood syndrome to learn of organizations' concern for their employees and contributions to the good of society.

Finally, business ethics researchers may wish to consider responsibility denial and Robin Hood syndrome in their future research. Greater understanding of how these personality characteristics are developed and changed, as well as their role in other computer ethics abuses, should prove fruitful.

ENDNOTES:

[1] The term, ethical decision making, refers to the process of coming to a decision involving ethics. The actual decision arrived at may be ethical or unethical.

[2] The term "moral" is interchangeable with "ethical" in this literature. Hence, this paper will use the term, ethical, since it is the term most appropriate in this study.

REFERENCES

Baum, R. J. (1989). Carts, Horses, and Consent: An Ethical Dilemma for Computer Networking Policy. In C. C. Gould (Ed.), *The Information Web: Ethical and Social Implications of Computer Networking* (pp. 87-100). Boulder, CO: Westview Press.

Bloombecker, B. (1990a). *Spectacular Computer Crimes*. Homewood, IL: Dow Jones-Irwin.

Bloombecker, B. (1990b). Needed: Binary Bar Mitzvahs and Computer Confirmations. *Computers & Society 20*(4), 20-21.

Business Software Alliance, Software & Information Industry Association (August, 1999), as reported in "Pirates Raid World Businesses," *Mobile Computing & Communications*, 24.

Conger, S., Loch, K. D., and Helft, B. L. (1995). Ethics and Information Technology Use: A Factor Analysis of Attitudes to Computer Use, *Information Systems Journal 5*(3), July, 161-184.

Eining, M. M. & Christensen, A. L. (1991). A Psycho-Social Model of Software Piracy: The Development and Test of a Model. In R. Dejoie, G. Fowler, and D. Paradice (Eds.), *Ethical Issues in Information Systems* (pp. 182-188). Boston, MA: Boyd & Fraser Publishing Company.

Forester, T. and Morrison, P. (1990). *Computer Ethics: Cautionary Tales and Ethical Dilemmas in Computing*. Cambridge, MA: The MIT Press.

Harrington, S. J. (1989). Why People Copy Software and Create Computer Viruses: Individual Characteristics or Situational Factors. *Information Resources Management Journal 2*(3), 28-37.

Harrington, S. J. (1995). The Anomaly of Other-Directedness: When Normally Ethical IS Personnel Are Unethical. *Computer Personnel, 16*(2), April, 3-11.

Henry, J. W. & Pierce, M. A. (1994). Computer Ethics: A Model of the Influences on the Individual's Ethical Decision Making. *Computer Personnel, 15*(3), October, 21-27.

Hunt, S. D. & Vitell, S. (1986). A General Theory of Marketing Ethics. *Journal of Macromarketing, 6*(1), 5-16.

Jones, T. M. (1991). Ethical Decision Making by Individuals in Organizations: An Issue-Contingent Model. *Academy of Management Review 16*(2), 366-395.

Kelman, H. C. & Hamilton, V. L. (1989). *Crimes of Obedience*. New Haven, CT: Yale University Press.

Kohlberg, L. and Candee, D. (1984). The Relationship of Moral Judgment to Moral Action. In W. M. Kurtines & J. L. Gewirtz (Eds.), *Morality, Moral Behavior, and Moral Development* (pp. 52-73). New York: John Wiley & Sons

Krauss, L. and MacGahan, A. (1979). *Computer Fraud and Countermeasures*, Prentice-Hall, Inc., Englewood Cliffs, NJ, pp. xi-xii.

Ladd, J. (1989). Computers and Moral Responsibility: A Framework for Ethical Analysis. In C. C. Gould (Ed.), *The Information Web: Ethical and Social Implications of Computer Networking* (pp. 207-227). Boulder, CO: Westview Press.

Nisan, M. (1984). Content and Structure in Moral Judgment: An Integrative View. In W. M. Kurtines & J. L. Gerwitz (Eds.), *Morality, moral behavior, and moral development* (pp. 208-224), New York: Wiley.

Parker, D. (1983). *Fighting Computer Crime*. New York: Charles Scribner's Sons.

Parker, D. (1989). *Computer Crime: Criminal Justice Resource Manual* (2nd ed.), National Institute of Justice, U. S. Department of Justice, pp. 7-9.

Perrolle, J. (1987). *Computers and Social Change: Information, Property, and Power*. Belmont, CA: Wadsworth Publishing.

President's Council on Integrity and Efficiency (1986). *Computers: Crimes, Clues, and Controls*. Washington, D.C.: U. S. Printing Office.

Rest, J. R. (1986). *Moral Development: Advances in Research and Theory*. New York: Praeger Publishers.

Samuelson, P. (1990). Computer Viruses and Worms: Wrong, Crime, or Both? In P. J. Denning (Ed.), *Computers Under Attack: Intruders, Worms, and Viruses* (pp. 479-485). New York: ACM Press.

Schwartz, S. H. (1973). Normative Explanations of Helping Behavior: A Critique, Proposal, and Empirical Test. *Journal of Experimental Social Psychology 9*, 349-364.

Schwartz, S. H. (1977). Normative Influences on Altruism. In L. Berkowitz (Ed.) *Advances in Experimental Social Psychology, Vol. 12* (pp. 221-279). New York: Academic Press.

Simpson, P. M., Banerjee, D. & Simpson, C. L. (1994). Softlifting: A Model of Motivating Factors. *Journal of Business Ethics, 13*, 431-438.

Snyder, M. & Swann, W. B. (1978). Behavioral Confirmation in Social Interaction: From Social Perception to Social Reality. *Journal of Experimental Social Psychology 14*, 148-162.

Staub, E. (1978). *Positive Social Behavior and Morality: Social and Personal Influences*. New York: Academic Press.

U. S. Department of Justice, National Institute of Justice (1989a). In J. T. McEwen (Ed.), *Dedicated Computer Crime Units*. Washington, DC: U. S. Government Printing Office.

U. S. Department of Justice, National Institute of Justice (1989b). In D. B. Parker (Ed.), *Computer Crime: Criminal Justice Resource Manual* (2nd ed.). Washington, DC: U. S. Government Printing Office.

Vitell, S. J. & Grove, S. J. (1987). Marketing Ethics and the Techniques of Neutralization. *Journal of Business Ethics, 6*, 433-438.

Watson, R. T. & Pitt, L. F. (1993). Determinants of Behavior Towards Ethical Issues in Personal Computing. *OMEGA International Journal of Management Science, 21*(4), 457-470.

Chapter 14

Social Issues in Electronic Commerce: Implications for Policy Makers

Anastasia Papazafeiropoulou and Athanasia Pouloudi
Brunel University, UK

INTRODUCTION

Policy implementation for electronic commerce is a complex process since policy makers, national governments in their majority, have to act in a fast changing environment. They need to balance special national demands with international cooperation (Papazafeiropoulou & Pouloudi, 2000). One of the areas that policy makers have to tackle is dealing with barriers that have been reported in the adoption of electric commerce today. These barriers are mostly derived from factors such as lack of awareness about the opportunities offered by electronic commerce as well as lack of trust to ward network security. Additionally the current legislative framework, drawn before the advent of electronic commerce, is perceived as outdated, thus impeding the expansion of online transactions. Policy makers, therefore, find it increasingly critical to update commerce legislation (Owens, 1999; Shim et al., 2000; the White House, 1999) and take other measures to facilitate the uptake of electronic commerce.

As the need for appropriate policy measures that support the information society is increasing, it is important to prevent a predominantly technical, commercial or legal approach that neglects the broader social issues related to policy making. To this end, this chapter examines social issues related to electronic commerce policy-making and is structured as follows. In the next section we present two fundamental social concerns that are related to policy making in electronic commerce: trust and digital democracy. In Section 3 we discuss these concerns in the light of different policy issues arising from the use of network technologies, and in Section 4 we present their implications for policy making in electronic commerce. The paper concludes with the importance of a holistic approach to policy making and suggestions for further research.

SOCIAL CONCERNS

The introduction of technologies such as the Internet in everyday life has resulted in a debate about its relative merits and disadvantages. Some of the social concerns are illustrated in the study conducted by the Stanford Institute for the Quantitative Study for Society (SIQSS, 2000) concerning the social implications of Internet use. The findings of the study indicate that the Internet is an "isolating technology" that could seriously damage the social fabric of communities as users interact physically with other people less. The social implications of the Internet can be witnessed in organizational processes, the nature of work, learning and education, innovation and competition, electronic democracy, privacy and surveillance (Dutton, 1996). This section considers the social concerns related to the use of Internet technologies by focusing on two of the most frequently discussed social issues in electronic commerce. These are trust, a social issue underlying the business use of the Internet, and digital democracy, a term underlying the use of Internet technology in the society as a whole. The following paragraphs consider each in detail.

Trust

Lack of trust in online transactions is one of the main reasons reported for the relatively low electronic commerce adoption today. Trust is a key issue and its existence among the business community and the end consumers will increase the willingness of trading partners to expand their electronic transactions (e.g., Hart & Saunders, 1997; Miles & Snow, 1992; Ratnasingham, 1998; Wilson, 1997). The low level of trust in electronic commerce can be attributed partly to the lack of face-to-face interaction between trading partners in conjunction with the general uncertainty of users in taking advantage of network technologies (Ratnasingham, 1998). According to Johnston (1999), there are a number of actions that can be taken to respond to user uncertainty. First, users should be educated about privacy and security issues. Second, the necessary legislation framework that protects trading partners must be developed. Third, the perceptions about technology as a tool that can threaten trust need to change to acknowledge that technology can also be applied for the users' protection, for example, through the effective use of encryption mechanisms.

Digital Democracy

Information and communication technologies offer opportunities for governments and citizens to be brought into closer dialogue; they also facilitate political organization and debate (Raab et al., 1996). However, the extent to which the information superhighway can fully enable citizens to participate in this emerging "digital democracy," has been heavily debated. First, at a conceptual level, our understanding of democracy is "as bounded in time as it is rooted in space" (Nguyen & Alexander, 1996, p. 120), which means that the term digital democracy is inherently problematic in "cyberspace." Importantly, there is a concern that if citizens are not able to have access to online services, because they do not have the means or the knowledge to do so, existing patterns of inequalities will be reinforced. The digital democracy is threatened by "information aristocracy" (Carter, 1997). In particular, there is evidence of a gender and race gap in the use of the Internet as well as differences for users with different levels of income and education (Hoffman & Novak, 1999; Kouzmin et al., 1999). While policy makers at an international level are concerned about access to electronic commerce, the burden falls mostly upon local authorities, which are responsible for the provision of access to network facilities through the use of public access centers, kiosks or tele-working centers. At a global level, the penetration of electronic commerce in developing countries is also an outstanding issue related to the "haves" and "have-nots" in cyberspace (e.g., Bhatnagar, 1997; Blanning et al., 1997; Clark & Lai, 1998; Kim & Hong, 1997). Easy global information access, however, is also problematic as it has been described as threatening both cultural identity and the regulatory sovereignty of the state, especially when used in less powerful economies (Shields, 1996). Finally, as privacy protection is a major concern in electronic commerce, there is a concern on whether "cyberspace" can promote democracy while protecting privacy. The free information flow of democracy and the users' need to control the flow of personal data can be seen as zero-sum alternatives that may (or may not) be balanced (Raab, 1997). This generates several policy dilemmas, which are reviewed in the following sections.

EMERGENT POLICY ISSUES

The Internet is the most popular means for the implementation of electronic commerce systems. Its fast expansion in the last decade was exceptional, forcing policy makers to speed up their efforts for its governance and regulation. The policy issues described in this section have to be addressed in order to facilitate the development of a safe and well-defined environment for electronic commerce, addressing the social concerns outlined in the previous section. These policy issues are presented following the six levels of Internet policy architecture including infrastructure, governance, security, privacy, content and commerce. These have been defined by the Global Internet Project (GIP), a group of senior executives from leading companies around the world (Patrick, 1999; www.gip.org). The second part of the section presents the dilemmas in addressing policy issues, leading on to a discussion of the implications for policy makers in the remainder of the chapter.

Policy Issues at Six Levels of Internet Policy

Infrastructure

The infrastructure level aims at addressing "the challenge of meeting the demand for reliable and scaleable access to the Internet" (Patrick, 1999, p. 106). The speed, the quality, the reliability and the cost of the networks used for online transactions are very important factors that can either boost or obstruct evolution of electronic commerce. One of the top priorities of governments is the support of the telecommunication industry so that it can offer better quality services in terms of speed, reliability, continuous access and interconnectivity between subnetworks (Patrick, 1999). The American government, for example, aims at the provision of online services to the majority of American households not only through desktop computers connecting to the Internet but also through devices such as television, cellular phones and portable digital assistants (US Department of Commerce, 1998). The liberalization of the telecommunication market is a relevant directive of the European Union (EC, 1997) and OECD (OECD, 1997b) to their member states. It demonstrates the intention of international policy making organizations to reduce the cost and improve the robustness of the telecommunication infrastructure worldwide.

In relation to the social concerns discussed in the previous section, policies that support the infrastructure level contribute towards better trust in terms of Internet performance. The availability of appropriate infrastructure and the capability to access it, however, as a prerequisite for the digital democracy, are contingent on the resources available within a particular region or country. Thus, global coverage is a major concern for policy makers today (Hudson, 1999). Within a national context, the quality of the telecommunication infrastructure in rural areas is particularly significant, when the accessibility to alternative means of obtaining information is very limited. Overall, as the role of the nation state declines in providing access to telecommunications networks, it may be up to independent bodies to support citizens gaining access to Internet-delivered services (Keenan & Trotter, 1999). At an international level, also, it may be up to independent bodies and international organizations to facilitate the development of Internet and technological infrastructure in developing countries. National governments also take initiatives to improve the take-up and use of information technologies but they do not always succeed (e.g., Walsham, 1999).

Governance

The Internet is characterized by its ability to expand without central governance. The Internet is the "place" where the free economy can blossom and this presents immense opportunities for electronic commerce. It is the intention of the policy makers at an international level to support industry leadership and self-regulation for electronic commerce (The White House, 1999; EC, 1997; OECD, 1997b). Specifically, there is a tendency to minimize government involvement and avoid unnecessary restrictions on electronic commerce.

However, as electronic commerce use becomes mature its international nature creates the need for global governance in certain areas. For example, several legal cases have been reported that involved Web site owners and consumers or other companies. The conflict usually derives from the lack of certainty about where a Web company is physically located and thus under which country's legal system the company works (Aalberts & Townsend, 1998). Taxation is a specific concern for companies that intend to invest in new technologies and for governments that want to control electronic commerce similarly to traditional commerce. There is a wide range of proposals concerning the administration of taxes in electronic commerce (Johnston, 1999; Owens, 1999). At one extreme there is the idea of absolute "tax-free" electronic commerce that has already been implemented for transactions taking place among US states, until February 1998 when the US public administration reaffirmed its commitment to making cyberspace a free-trade zone (Negroponte, 1999). At the other end there are proposals for introduction of special new taxes for electronic commerce. OECD (1997a) proposes an intermediate solution, directing its members to apply existing tax principles in electronic transactions. OECD, in co-operation with the European Union, the World's Customs Organization and the business community, has defined a set of framework conditions to govern the taxation of electronic commerce. These are neutrality, effectiveness and fairness, certainty and simplicity, efficiency and flexibility, factors that are naturally important to traditional commerce as well. Thus it is necessary to define the "rules" that govern electronic commerce and ensure that regulations can be enforced.

Overall, the governance level of Internet policy presents a challenge for national policy makers as they realize it is difficult, if not impossible, to control electronic transactions. Also, it is debatable what is within a specific jurisdiction or how "net-laws" will be enforced or who will pay for enforcement (Shim et al., 2000). Additionally, policy makers are also keen to promote electronic commerce with minimal intervention, as they want to attract investors that will contribute to economic growth. North American countries, the European Union and Japan for example have realized that it is in their best interest to collaborate in order to create market conditions of trust. However, the interests of specific countries may at times prevail, and the compromises reached may be at a cost for digital democracy. A characteristic example is the difference between European and American provisions for personal data protection and its impact on electronic transactions between the two areas. This issue is addressed in further detail at the security and privacy levels in the next paragraphs.

Security

Network security and especially Web security is one of the most sensitive issues identified in the electronic commerce literature (e.g., Crocker, 1996; Kosiur, 1997; Liddy, 1996). A recent survey of Australian firms (Dinnie, 1999), "among the world's earliest adopters" of electronic commerce, reports that network security is a continuing concern and companies are more concerned about external threats. The survey reports that "16% of firms have suffered, or believe they may have suffered, at least one break-in via the Internet" (p. 112). Despite their perceptions of external threats, however, 30% of businesses admitted that their organization had no formal information security policy. More generally, the anxiety about security is expected to increase in coming years as Web-based applications are increasingly used for financial transactions. As the number of computers, networks, data and information multiply every day, the need for better security practices that protect information systems from malicious attacks and at the same time preserve the civil liberties will increase in the future (Hurley, 1999).

Cryptography is put forward as a powerful technological solution to network fraud. At an international level it can be applied with the collaboration of governments, the business community and trusted third parties (Denning, 1996). The required use of public and private keys in cryptography methods raises several public policy issues surrounding the encryption of data and who should hold the keys that unlock the encrypted information (Patrick, 1999; Pouloudi, 1997). Policy makers can play an important role in the implementation of a security policy, acting as trusted third parties or defining the legal framework for such organizations (Froomkin, 1996). There are multiple models concerning the role of governments in security policy. At one extreme, public authorities may have ultimate access to information and, at the other, they may leave the responsibility for security of the data to the information owner (Patrick, 1999). What seems to be urgently required today is better education and awareness of security of information systems and good security practices for companies and individuals (Hurley, 1999).

Privacy

Computer technologies like the Internet facilitate the exchange of personal information that can be collected, aggregated and sold across the world. As companies can easily take advantage of personal information that becomes accessible on information networks, e.g., through direct marketing (Wang et al., 1998), several issues are at stake. The most important concern is whether information is collected, aggregated or sold with the individual's explicit concern. There are several private organizations (Better Business Bureau onLine, BBBOnLine; World Wide Web Consortium, W3C; TRUSTe) that try to address the issue by giving a privacy "seal" to Web sites that are fulfilling some set criteria of privacy protection. These include the responsibility to make visitors to Web sites aware of what data is collected and giving them choice about making this data available to third parties. The TRUSTe white paper (http://www.truste.org/about/about_wp.html) also emphasizes that web sites bearing their privacy seal "must provide reasonable security to protect the data that is collected." Security is seen as the technological aspect of the broader social issues that are related to privacy.

Privacy is particularly important for the protection of sensitive personal data such as medical records, credit records, government data and personal data about children. The US government has taken an untied regulatory approach to protect such information. In other words the aim is to enable Internet users to choose for themselves what level of privacy protection they want (Nelson, 1999). In Europe, in contrast, data protection is stricter and has been articulated at a pan-European level (Allaert & Barber, 1998). In the United States, the EU directive (EC, 1995) has been perceived as being overprotecting for European companies, raising barriers to the free exchange of electronic data between Europe and other countries (Swire & Litan, 1998). Indeed, the European directive on data protection challenged electronic transactions and data exchanges internationally, as it banned the export of personal data from the EU to those countries without strict federal data protection laws. This included the US and resulted in severe trade disputes at an international level, which have been resolved recently with the Safe Harbor Privacy Arrangement. This is a mechanism with which, through an exchange of documents, EU is able to certify that participating US companies meet the EU requirements for adequate privacy protection. Participation in the safe harbor is voluntarily. Privacy advocates, however, argue that privacy is a profound and fundamental concept, hence "it merits extra-ordinary measures of protection and overt support" (Introna, 1997, p. 259).

The political nature of privacy is also evident within national boundaries, in particular in terms of the power that national regulators have: "what we should fear is the growth of government databases" (Singleton, 1998). Privacy therefore clearly raises social concerns in terms of trust, digital democracy as well as employment, particularly in relation to the rights of employers to access or monitor personal information of their employees (ranging from email messages to medical records), often without their explicit consent or even their knowledge. Finally, the difficulties of updating databases and business processes and the challenges to comply at a technical level when using some contemporary information technologies (Lycett & Pouloudi, 2002) signify that privacy protection remains a challenge for policy makers.

Content

As electronic commerce is an international phenomenon it is impossible for policy makers to control the content of the information transferred online. While the exposure to all this information can be beneficial, for example, expanding people's learning horizons (Forcheri et al., 2000), governments and citizens are concerned about the publication of offensive material (Nelson, 1999). As the complaints from parents and educators about the influence of the Internet on children become more frequent, there are several civil liberties organizations devoted to protecting users from exposure to inappropriate online material. Such groups include the Electronic Frontier Foundation (EFF), which supports legal and legislative action to protect the civil liberties of online users, and the Computer Professionals for Social Responsibility (CPSR), which aims to protect privacy and civil liberties. The World Wide Web Consortium (W3C) has developed a technical platform that allows user-defined, customized access to the Internet (Patrick, 1999; www.w3.organisation/PICS) and has enabled the creation of rating services and filtering software for use by concerned parents. While the need for filtering of some information is generally considered as appropriate, there are also attempts at censorship. For example certain Asian countries place restrictions on the use of the Internet. The use of censorship on the information highway is debatable, both in terms of its technological feasibility but also in terms of its moral foundation (Ebbs & Rheingold, 1997).

Another content-related issues in electronic commerce is the protection of copyright and intellectual property rights. The essence of copyright is to prevent the unauthorized copying, but works stored in a digital format can easily be copied or altered, while they can also be transmitted speedily through electronic networks (Brett, 1999). The practical problems that owners of digital data face are very important for governments trying to apply or extend existing copyright laws to digital means. At an international level the World Intellectual Property Organization (WIPO) facilitates the protection of property rights. According to its general director, Dr. Kamil Idris, the organization's aim is to ensure that "expertise is provided when laws or systems need upgrading to take into account novel areas of invention (such as providing protection for the fruits of genetic research) or of medium (such as the Internet)." As with other policy issues, intellectual property involves multiple stakeholders with different interests (Radcliffe, 1999), which makes it difficult to resolve at a global level.

Underlying the discussion in terms of content are also issues of trust, in terms of access to "suitable" material but also in terms of authenticity, and issues related to the concept digital democracy, depending on who, if any, decides what constitutes "suitable" material.

Commerce

Electronic commerce is at the top of the policy architecture pyramid of the Global Internet Project, as it is perceived to be a critical factor driving the growth of the Internet. Although electronic commerce has revolutionized the way of conducting business, it is still a business activity that has to conform to certain rules and work under specific standards (Negroponte, 1999). The European Union was the first official body that considered a supranational policy on electronic commerce, in its effort to advance the integration process and to create a single market (Mc Gowan, 1998). However, there are several organizations working at a supranational level trying to enable global seamless communication such as the International Organization for Standardization (ISO) and the World Trade Organization (WTO). This is because standardization is recognized as an important issue in electronic commerce, since the establishment of EDI applications (e.g., Chatfield & Bjorn-Andersen, 1998; Faltch, 1998; Sokol, 1995; Tan, 1998). Standardization however can be problematic, as it needs to balance multiple interests in an area where competition has international dimensions and differs considerably from traditional commerce. The extent to which certain stakeholders are privileged has an impact on the role of electronic commerce in facilitating the digital democracy. The importance of trust at this level cannot be understated since, as discussed in section 2, it is one of the main reasons why electronic commerce has not reached its current potential.

The discussion in the previous five levels of the policy architecture demonstrates that issues of trust are relevant at all levels and indeed underpin the development and use of electronic commerce. The problem is that most of these policy issues are related to social concerns and cannot be easily resolved as they bring about conflicts among stakeholder groups and policy dilemmas. These dilemmas are discussed in detail below in the context of electronic commerce policy-making.

Dilemmas in Addressing Policy Issues

Previous research has argued that the policy objective of promoting deregulation and competition is in conflict with other policy priorities, in particular the desire to provide open networks and open access and the aspiration to provide universal service to citizens (Graham, 1995). As electronic commerce expands, the dilemmas for the stakeholders of the information society increase. The review of policy issues at different levels in the previous section has revealed some of the dilemmas that policy makers face today:

- Should governments give priority to the protection of national identity and language or to international compliance?
- Should they promote their own interests or provide assistance to developing countries?
- Is governance about protection or restriction? (For example, at an individual level: is censorship desirable? At a business level: is taxation desirable?)
- Where should priority be given: to the protection of personal data or to competitiveness (to the extent that the free exchange of information and personal data supports electronic transactions and business practices)?
- What is more important, data and intellectual property protection or the free exchange of ideas and data?

These dilemmas relate to the appropriate use of regulation, although in some cases policy makers may have little choice as only some options are realistic (e.g., the Internet is used even though the legal context is unstable). Thus, one important observation is that some dilemmas may no longer be a matter of choice, particularly for less powerful stakeholders, such as individuals, or governments of developing countries. A further observation is that in many cases these dilemmas imply a conflict between the commercial and social interests of various stakeholder groups. However, it is very difficult to draw some general conclusions about when either interest is at stake. Research in management (e.g., Pettigrew, 1985) and information systems (e.g., Walsham, 1993) as well as in law studies (e.g., as evident in the importance of case law) has stressed the importance of context. However, in "cyberspace" the context, whether temporal or spatial, is elusive, making policy making for electronic commerce more challenging. In view of these issues, the following section presents implications for policy makers, with emphasis on the policies that are relevant at the business and the societal level.

Table 1: Target groups of an electronic commerce strategy

Policies	Individuals– Societal level	Companies- Business level
Knowledge building	✔	✔
Knowledge deployment	✔	✔
Subsidy		✔
Mobilization		✔
Innovation directive		✔
Standard/regulation setting		✔

IMPLICATIONS FOR POLICY MAKERS

The challenge that policy makers face today in order to implement an efficient electronic commerce policy while addressing the dilemmas outlined above is twofold. Firstly, they need to provide the business community with a robust technical infrastructure and an efficient legislation framework. Secondly, they need to accommodate the social concerns rising from the use of electronic commerce, in order to create a "digital literate" society that will fully exploit the technology at hand while preserving their social interests and cultural identities.

A very important aspect of a national electronic commerce strategy is diffusion of knowledge about the to business and society at large. Damsgaard and Lyytinen (1998) use, in their analysis on the diffusion of EDI (business-to-business electronic commerce), six government strategies defined by King et al. (1994). These are knowledge building, knowledge deployment, subsidy, mobilization, innovation directive, and standard setting. We extend these strategies for the diffusion for electronic commerce, where apart from business individuals are also the targets of the government intervention. Thus, a grid can be created (see Table 1) with the combination of these strategies and their target groups (business, society).

Companies are usually the direct beneficiaries of electronic commerce policies. This is why all the diffusion strategies are applicable (see far right column in Table 1). Policy makers try to persuade enterprises to invest in new technologies and take advantage of the opportunities the new means can offer. The governments may use a great number of the strategies to influence companies and help them in the implementation of electronic commerce technologies and practices. Companies can first be made aware of the new technologies (knowledge deployment), receive financial support for investing in new technologies (subsidy), be encouraged to use technology in the "best way" (mobilization), be provided with examples of electronic commerce use (innovation directive) and finally follow standards (regulation setting). This part of the electronic commerce diffusion practice is related to technical and commercial aspects, which as we will explain in the next paragraph can be conflicting with social issues.

Individuals acting as consumers (such customers of virtual stores) or citizens (as users of online government services) are in need of information. Governments can use traditional means such as the media to make their wide audience aware of the usefulness of the new medium and build confidence in electronic commerce transactions. Knowledge deployment and mobilization are the strategies that can best fit government's intention to create awareness about electronic commerce, as well as about the rights of individuals in this new environment. Issues such as awareness about privacy protection and trust towards electronic means should be considered by policy makers when they apply knowledge building and deployment practices. The education of the public, on one hand, can help the electronic commerce marketplace to reach a critical mass of users. On the other hand, a "digital literate" society can use electronic means to perform "electronic activism" and express disappointment about business practices (see, for example Badaracco & Useem, 1997). Additionally, they might refuse the exchange of personal data through electronic means, although this is a practice that is very useful to companies for marketing purposes. Thus, regulators should balance the needs of the business community with the social concerns related to the use of electronic means. It is expected that when the social issues such as trust and digital democracy are addressed satisfactorily, electronic commerce is more likely to become the predominant business practice.

The "education" of individuals within the business environment (business level) is essential. In this field the help of professional bodies such as chambers of commerce and trade associations is essential. While most of policy research concentrates on the role of governments or international organizations, the role of players, such as trade associations, that can act as policy intermediaries is very important: They have knowledge of the local context and thus can complement the general national or international policies. As discussed earlier in the paper, other policy intermediaries that become increasingly involved in policy issues in the information society include independent private organizations as well as civil liberties and professional groups who wish to promote the interest of a particular group or the net-citizens at large. Schools and universities also face pressures to support the "workforce of the future" and try to promote the use of information and communication technologies, thus contributing to knowledge building and deployment strategies. Finally, the Internet empowers individuals to draw their own policies at a micro-level, e.g., choosing as parents which Internet sites they allow their children to access, deciding whether to make their personal information available and so on. While the Internet enables people as citizens and consumers to take action (e.g., Badaracco & Useem, 1997), people are not necessarily aware of the opportunities and risks of cyberspace or they may not have the power and access to make a difference, hence the importance of knowledge building and deployment strategies. Policy makers, whether local or national, government or private, need to recognize the prevalence and importance of social issues and encourage the debate for appropriate policy making among stakeholders.

CONCLUSIONS

Policy makers have recognized the viability of electronic commerce and the opportunities it offers for business and citizens. While several ethical and security issues arise from the use of the new technologies, there is a general consensus that the benefits are substantial and justify the investment in electronic commerce. There are several efforts in this direction by policy makers at a national and international level. The paper has argued that technology alone is not sufficient for the successful implementation of complex electronic commerce strategies, but the examination of social and political issues is crucial for a holistic approach on the subject. Indeed there are several dilemmas related to policy issues, making the role of the policy makers critical. We considered a general framework for policy making that could be used at a national or international level as a starting point for considering social issues in the context of electronic commerce strategies.

Further research in the area may include the investigation of electronic commerce policies implemented in different national settings and social environments since, in practice, different countries have different priorities. The case of developing countries would be of particular interest as technical infrastructure and stakeholder awareness and involvement can be substantially different. Research also needs to be continued in specific areas that are affected by the extensive use of electronic commerce. Because of their social importance, of particular interest are the areas of health and education where issues of Internet use and electronic commerce become increasingly relevant (e.g., through tele-health or distance learning applications). A study of alternative national policies in these areas can lead to an informative debate about the underlying assumptions concerning the duties and social responsibility of policy makers towards different stakeholder groups.

ACKNOWLEDGMENTS

The financial support of EPSRC (grant GR/N03242) is gratefully acknowledged.

REFERENCES

Aalberts, R., and Townsend, A. (1998). The threat of long-arm jurisdiction to electronic commerce. *Communications of the ACM*, 41(12), 15-20.

Allaert, F. A. and Barber, B. (1998). Some systems implications of EU data protection directive. *European Journal of Information Systems*, 7(1), 1-4.

Badaracco, J. L., Jr. and Useem, J. V. (1997). The Internet, Intel and the vigilante stakeholder. *Business Ethics: A European Review*, 6(1), 18-29.

Bhatnagar, S. (1997). Electronic commerce in India: The untapped potential. *Electronic Markets*, 7(2), 22-24.

Blanning, R., Bui, T. and Tan, M. (1997). National information infrastructure in Pacific Asia. *Decision Support Systems*, 21, 215-227.

Brett, H. (1999). Copyright in a digital age. In Leer, A. (Ed.), *Masters of the Wired World*, 162-171. London: Financial Times Pitman Publishing.

Carter, D. (1997). Digital democracy or information aristocracy economic regeneration and the information economy. In Loader, B. (Ed.), *The Governance of Cyberspace*, 136-152. Routledge, London.

Chatfield, A. and Bjorn-Andersen, N. (1998). Reengineering with EDI. A Trojan horse in circumventing non-tariff barriers to trade. In Andersen, K. V. (Ed.), *EDI and Data Networking in the Public Sector*, 155-172. Kluwer Academic Publishers.

Clark, J., and Lai, V. (1998). Internet comes to Morocco. *Communications of the ACM*, 41(2), 21-23.

Damsgaard, J. and Lyytinen, K. (1998). Governmental intervention in the diffusion of EDI: Goals and conflicts. In Andersen, K. V. (Ed.), *EDI and Data Networking in the Public Sector*, 13-41. Kluwer Academic Publishers.

Daniel, J. (1999). The rise of the mega-university. In Leer, A. (Ed.), *Masters of the Wired World*, 333-342. London: Financial Times Pitman Publishing.

Dinnie, G. (1999). The second annual global information security survey. *Information Management & Computer Security*, 7(3), 112-120.

Doukidis, G., Poulymenakou, A., Terpsidis, I., Themistocleous, M. and Miliotis, P. (1998). *The Impact of the Development of Electronic Commerce on the Employment Situation in European Commerce*. Greece: Athens University of Economics and Business.

Dutton, W. H. (Ed.). (1996). *Information and Communication Technologies: Visions and Realities*. Oxford, UK: Oxford University Press.

Ebbs, G., & Rheingold, H. (1997). Censorship on the information highway. *Internet Research: Electronic Networking Applications and Policy*, 7(1), 59-60.

European Commission. (1995). Directive 95/46/EC of the European Parliament and the Council of 24 October 1995 on the protection of individuals with regard to the processing of personal data and on the free movement of such data. *Official Journal of the European Communities*, November, L281, 31.

European Commission. (1997). *A European Initiative in Electronic Commerce* (COM (97) 157 final). Brussels, Belgium.

Faltch, M. (1998). EDI in the public sector: Building on lessons from the private sector. In Andersen, K. V. (Ed.), *EDI and Data Networking in the Public Sector*, Kluwer Academic Publishers.

Forcheri, P., Molfino, M. T. and Quarati, A. (2000). ICT driven individual learning: new opportunities and perspectives. *Educational Technology & Society*, 3(1), 51-61.

Froomkin, A. (1996). The essential role of trusted third parties in electronic commerce. In Kalakota, R. and Whinston, A. (Eds.), *Readings in Electronic Commerce*, Addison-Wesley.

Graham, A. (1995). Public policy and the information superhighway: the scope for strategic intervention, co-ordination and top-slicing. In Collins, R. and Purnell, J. (Eds.), *Managing the Information Society*, 30-44. London: Institute for Public Policy Research.

Hart, P. and Saunders, C. (1997). Power and trust critical factors in the adoption and use of electronic data interchange. *Organization Science*, 8(1), 23-41.

Heldrich Center for Workforce Development. (2000). *Work Trends Survey, Nothing but Net: American Workers and the Information Economy*, 10 February.

Hoffman, D. and Novak, T. (1999). The evolution of the digital divide: Examining the relationship of race to Internet access and usage over time. *Conference on Understanding the Digital Economy: Data, Tools and Research*, May. USA: Washington.

Hudson, H. (1999). Access to the digital economy: Issues in rural and developing regions. *Conference on Understanding the Digital Economy: Data, Tools and Research*, 25-26 May. USA: Washington.

Hurley, D. (1999). Security and privacy laws. The showstoppers of the Global Information Society. In Leer, A. (Ed.), *Masters of the Wired World*, 247-260. London: Financial Times Pitman Publishing.

Introna, L. D. (1997). Privacy and the computer: Why we need privacy in the information society. *Metaphilosophy*, 28(3), 259-275.

Johnston, D. (1999). Global electronic commerce-realizing the potential. In Leer, A. (Ed.), *Masters of the Wired World*, 228-237. London: Financial Times Pitman Publishing.

Keenan, T. P. and Trotter, D. M. (1999). The changing role of community networks in providing citizen access to the Internet. *Internet Research: Electronic Networking Applications and Policy*, 9(2), 100-108.

Kim, E. and Hong, P. (1997). The government's role in diffusion of EC in Korea. *Electronic Markets*, 7(2), 6-8.

King, J., Gurbaxani, V., Kraemer, K., McFarlan, F., Raman, F. and Yap, F. W. (1994). Institutional factors in information technology innovation. *Information Systems Research*, 5(2), 139-169.

Kouzmin, A., Korac-Kakabadse, N. and Korac-Kakabadse, A. (1999). Globalization and information technology: Vanishing social contracts, the "pink collar" workforce and public policy challenges. *Women in Management Review*, 14(6), 230-251.

Lycett, M. G. and Pouloudi, A. (2002). Component-based development: issues of data protection. In Dhillon, G. (Ed.), *Social Responsibility in the Information Age*. Hershey, USA: Idea Group Publishing.

Martin, J. (1999). Building the cyber-corporation. In Leer, A. (Ed.), *Masters of the Wired World*, 324-332. London: Financial Times Pitman Publishing.

Mc Gowan, L. (1998). Protecting competition in a global market: The pursuit of an international competition policy. *European Business Review*, 98(6), 382-339.

Miles, R. and Snow, C. (1992). Causes of failure in network organizations. *California Management Review*, Summer, 53-72.

Murison-Bowie, S. (1999). Forms and functions of digital content in education. In Leer, A. (Ed.), *Masters of the Wired World*, 142-151. London: Financial Times Pitman Publishing.

Negroponte, N. (1999). Being digital in the wired world. In Leer, A. (Ed.), *Masters of the Wired World*, 386-394. London: Financial Times Pitman Publishing.

Nelson, M. (1999). Politics and policy-making in the electronic marketplace. In Leer, A. (Ed.), *Masters of the Wired World*, 261-269. London: Financial Times Pitman Publishing.

Nguyen, D. T. and Alexander, J. (1996). The coming of cyberspacetime and the end of the polity. In Shields, R. (Ed.), *Cultures of Internet: Virtual Spaces, Real Histories, Living Bodies*, 99-124. London: Sage.

Organization for Economic Co-operation and Development. (1997a). *The Communication Revolution and Global Commerce: Implications for Tax Policy and Administration*.

Organization for Economic Co-operation and Development. (1997b). Global Information infrastructure-Global information society (GII-GIS). Policy requirements.

Owens, J. (1999). Electronic commerce: Taxing times. In Leer, A. (Ed.), *Masters of the Wired World*, 286-295. London: Financial Times Pitman Publishing.

Papazafeiropoulou, A., and Pouloudi, A. (2000). The government's role in improving electronic commerce adoption. In Hansen et al., (Eds.), *Proceedings of the European Conference on Information Systems 2000*, July, 1, 709-716. Austria: Vienna.

Patrick, J. (1999). The opportunity and the challenge to sustain rapid Internet growth. In Leer, A. (Ed.), *Masters of the Wired World*, 105-112. London: Financial Times Pitman Publishing.

Pettigrew, A. M. (1985). Contextualist research and the study of organizational change processes. In Mumford, E., Hirschheim, R., Fitzgerald, G. and Wood-Harper, T. (Eds.), *Research Methods in Information Systems*, 53-78. Amsterdam: Elsevier Science Publishers, North-Holland.

Raab, C. (1997). Privacy, democracy, information. In Loader, B. (Ed.), *The Governance of Cyberspace*, 155-174. London:Routledge.

Raab, C., Bellamy, C., Taylor, J., Dutton, W. H. and Peltu, M. (1996). The information polity: Electronic democracy, privacy, and surveillance. In Dutton, W. H. (Eds.), *Information and Communication Technologies*, 283-299. Oxford, UK: Oxford University Press.

Radcliffe, M. (1999). Intellectual property and the global information infrastructure. In Leer, A. (Ed.), *Masters of the Wired World*, 105-112. London: Financial Times Pitman Publishing.

Ratnasingham, P. (1998). The importance of trust in electronic commerce. *Internet Research: Electronic Networking Applications and Policy*, 8(4), 313-321.

Shade, L. R. (1996). Is there free speech on the Net? Censorship in the global information infrastructure. In Shields, R. (Ed.), *Cultures of Internet: Virtual Spaces, Real Histories, Living Bodies*. London: Sage.

Shields, R. (Ed.). (1996). *Cultures of Internet: Virtual Spaces, Real Histories, Living Bodies*. London: Sage.

Shim, J. P., Simkin, M. G. and Bartlett, G. W. (2000). NetLaw. *Communications of the Association for Information Systems*, 4(4).

Singleton, S. (1998). Privacy as censorship: A skeptical view of proposals to regulate privacy in the private sector (Cato Policy Analysis No. 295).

Sokol, P. (1995). *From EDI to Electronic Commerce*. McGraw-Hill.

Stanford Institute for the Quantitative Study for Society. (2000). *Internet and Society*, February.

Swire, P. P., and Litan, R. E. (1998). None of your business. *World Data Flows, Electronic Commerce, and the European Privacy Directive*. Washington DC, USA: Brookings Institution Press.

Tan, M. (1998). Government and private sector perspective of EDI: The case of TradeNet EDI and Data Networking in the Public Sector, Andersen, K. V. (Ed.), 131-153. Kluwer Academic Publishers.

US Department of Commerce. (1998). *The Emerging Digital Economy*. Washington, USA: US Department of Commerce.

Walsham, G. (1993). *Interpreting Information Systems in Organizations*. Chichester: Wiley.

Walsham, G. (1999). GIS for district-level administration in India: problems and opportunities. *MIS Quarterly*, 23(1), 39-66.

Wang, H., Lee, M. K. O. and Wang, C. (1998). Consumer privacy concerns about internet marketing. *Communications of the ACM*, 41(3), 63-70.

The White House. (1999). *Facilitating the Growth of Electronic Commerce*. Washington DC, USA: The White House.

Wilson, S. (1997). Certificates and trust in electronic commerce. *Information Management & Computer Security*, 5(5), 175-181.

Chapter 15

Kierkegaard and the Internet: The Role and Formation of Community in Education

Andrew Ward
University of Minnesota, USA

Brian Prosser
Fordham University, Bronx, USA

In the last decade of the twentieth century, with the advent of computers networked through Internet Service Providers and the declining cost of such computers, the traditional topography of secondary and post-secondary education has begun to change. Where before students were required to travel to a geographically central location in order to receive instruction, this is often no longer the case. In this connection, Todd Oppenheimer writes in *The Atlantic Monthly* that one of the principal arguments used to justify increasing the presence of computer technology in educational settings is that "[W]ork with computers – particularly using the Internet – brings students valuable connections with teachers, other schools and students, and a wide network of professionals around the globe."[1]

This shift from the traditional to the "virtual" classroom[2] has been welcomed by many. As Gary Goettling writes, "[D]istance learning is offered by hundreds, if not thousands, of colleges and universities around the world, along with a rapidly growing number of corporate and private entities."[3] Goettling's statement echoes an earlier claim by the University of Idaho School of Engineering that one of the advantages of using computers in distance education is that they "increase access. Local, regional, and national networks link resources and individuals, wherever they might be."[4]

Previously Published in *Challenges of Information Technology Management in the 21st Century*, edited by Mehdi Khosrow-Pour,

It is, though, all too easy to fail to reflect critically on the broader ramifications of Internet use. For example, a 1998 study conducted at Carnegie Mellon concludes that "[G]reater use of the Internet was associated with small, but statistically significant declines in social involvement as measured by communication within the family and the size of people's local social networks, and with increases in loneliness ... [and] with increases in depression."[5] The study goes on to say that the paradox of the Internet is that it is "a social technology used for communication with individuals and groups, but it is associated with declines in social involvement and the psychological well-being that goes with social involvement."[6] At its worse then, rather than the Internet providing a tool for the creation of an interactive environment, we have a situation in which, as Fred Moody, in a commentary for ABC News puts it, "[A]ny time we go online, we are replacing direct human contact ...with an arid, indirect, stilted form of contact with strangers."[7]

While the Carnegie Mellon study is provisional, it at least suggests that it is prudent to reflect upon how the use of the Internet in education might affect both students and instructors.[8] With this in mind, the purpose of this paper to is to examine whether it is truly the case that students are more closely linked together as a "community" of learners through the use of virtual classrooms. In this connection, University of California-Berkeley Professor Hubert Dreyfus has made use of the writings of one pre-digital skeptic, the 19th century philosopher Søren Kierkegaard, to explore potential limitations in using the Internet as a means for establishing social commitments.[9] In what follows we use Dreyfus' exploration in looking more closely at problems accompanying use of the Internet as a tool for creating "distributed communities" of teachers and learners.

Dreyfus argues that Kierkegaard's writings show concern about a diminishing ability to discern "quality" or meaningful information. As Kierkegaard saw it, the use of technologies to disseminate information distorts our relationship to that information in ways that foster an ability to ignore the potential "meaningfulness" of particular pieces of information. For Kierkegaard, it was the technology embodied in populist newspapers that was of the greatest concern. Dreyfus contends, however, that Kierkegaard's concern is even more appropriate in characterizing and evaluating on-line behavior. For Kierkegaard there is a link between the power of our information technologies to uninhibitedly disseminate information and a concomitant desire of participants to "transcend the local, personal involvement"[10] of information. It is this disembodiedness and dislocatedness (*omni*locatedness) of cyberspace that makes it an attractive replacement for the classroom. Placing educational materials in an omnilocated state makes them available to everyone who has the technological means to "transcend the local".

However, before we can approach the question of *how* cyberspace may be used to meet the goals of quality educational communication, there is the question of *whether* this can be adequately accomplished at all. In this connection, Kierkegaard writes that "all mankind's great inventions (railroads, telegraph, etc.) tend to develop and encourage windbaggery."[11] Further, he writes that this tendency "is continually in the direction of perfecting the means of communication so that the communication of nonsense can spread farther and farther."[12] As Dreyfus points out, Kierkegaard would see our on-line behaviors as a continuation of this trajectory.

The problem from Kierkegaard's point of view is not simply that technological development represents a trajectory toward "windbaggery," but that it is a trajectory *away from* interpersonal connectivity. "[P]ersonality," he writes, "has been abolished.... All communication is impersonal - and here in particular are the two most dreadful calamities which really are the principle powers of impersonality - [mass communication] and anonymity."[13] From Kierkegaard's perspective, these two "calamities" are linked. Technologically mediated communications run an increasing risk of attenuating interpersonal connectivity by transforming relationships to information in ways that promote anonymity and impersonality. By contrast, genuine communication is meant to be face-to-face communication between real, embodied individuals. As Kierkegaard puts it, "God really intended that a person should speak individually with his neighbor and at most with several neighbors."[14]

Using Kierkegaard's insights as a starting point, there are at least two ways in which technologically mediated communications compromise interpersonal connectivity. First, technologically mediated communication transforms a participant's relationship to the information that is exchanged in communication. Second, technologically mediated communication transforms our perception of persons participating in the communication activity. Regarding the first transformation, Sherry Turkle offers this anecdote:

> Peter, a twenty-eight-year-old lecturer in comparative literature, thought he was in love with a MUDding partner who played Beatrice to his Dante (their characters' names). Their relationship was intellectual, emotionally supportive, and erotic. Their virtual sex life was rich and fulfilling... Peter flew from North Carolina to Oregon to meet the woman behind Beatrice and returned home crushed. "[On the MUD] [said Peter] I saw in her what I wanted to see. Real life gave me too much information."[15]

While it may be hyperbole to claim that "the medium *is* the message," what such cases exemplify is the important role that the medium does play in communication. It shows that the medium of information exchange affects the beliefs, feelings and attitudes of those to whom the information is conveyed.

Regarding the second transformation, consider situations that reflect a change in self-perception stemming from the use of different communications media. For example, this paper's co-authors have had students say that they are more comfortable asking questions by e-mail or voice-mail than in person. When asked why, the gist of a typical answer is that e-mail and voice-mail offer a higher degree of anonymity. Although the Internet "has been praised as superior to television and other "passive" media because it allows users to choose the kind of information they want to receive, and often, to respond actively to it in the form of e-mail exchange,"[16] it also allows the student to avoid interacting with the instructor in face-to-face situations, and so encourages anonymity.[17]

We can apply these examples to Kierkegaard and see that there is a troubling possibility that the types of transformations affected by information technologies are ones that do diminish a healthy sense of "interpersonal connectivity." For example, with respect to anonymity, Peter Danielson writes that "[A]nonymous messages are especially disruptive of social cooperation. They undercut the recognition of participants that is the basis of reciprocity and discrimination."[18] Moreover, such examples also suggest that in computer mediated communication there always remains a sense that we are, to some extent, talking to "a box" rather than another human being. With "the box" standing between us, the reality of the situation is altered just enough, and we are able to feel just enough detachment and anonymity that the rules which bind us in embodied, face-to-face interaction are diminished.

A pessimism like Kierkegaard's rests in the belief that the contexts of embodied interaction create a particular connectivity essential to communication. Herein lies the depth of the problem. It *seems* that technology, because it permits a kind of omnilocatedness of individuals, should enhance interpersonal connectedness. If Kierkegaard is right though, rather than enhancing the possibilities of conversation and community building, the non-linear and non-hierarchical character of cyberspace is depersonalizing, and so antithetical to both possibilities.[19] By dis-locating the locus and structure of conversation, the Internet stymies and breaks down interpersonal connectedness. In this connection, the experiences of Ron Barnette are especially revealing. In 1994 Barnette launched his "Philosophy in Cyberspace" course – PHICYBER – "accessible on-line twenty-four hours a day, seven days a week, for the ten-week term."[20] The class began with twenty-one participants, but soon grew to one hundred eleven participants "from eleven countries representing five continents."[21] While PHICYBER exemplifies the ideal of an educational "community," informationally linked and geographically distributed, the cost is a distinctive kind of isolation for the participants. As Barnette writes:

> There are no voices or accents, no noises, nor distinctions based on gender, race, ethnicity or age. Only ideas, and ideas about ideas, for-

> mulated, written, and rewritten, expressed and revisited. In fact, the ongoing discussion in the class *is* the set of ideas expressed. A participant becomes, in a sense, a Platonist in cyberspace, instantiated by material objects and electricity![22]

It is this "post-modern" effect of dis-locatedness that leads to feelings of isolation as well as feelings that the world apart from cyberspace is somehow "too real."[23] At the same time though, the dis-locatedness engendered by Internet mediated communication also leads to a fracturing of the self.[24] Here it is interesting to note that Gary Shank and Donald Cunningham, for all their apparent support of using the Internet in community building, claim that in an Internet culture, as opposed to an oral culture, "[P]ersonal identities are less important…"[25]

If our broadly Kierkegaardian remarks are on track, then insofar as interpersonal connectivity is an essential component of quality education, its attenuation bodes ill for computer-mediated instruction. Neil Postman's remarks in the PBS documentary "net.learning" certainly typify such concerns: "[N]othing," he says, "can replace the bond of a teacher and a student who are physically together."[26] Contrary to Postman's pessimism, our conclusion here is *not* to say that there is no room for computer mediation in education. We should remember that the main point for Kierkegaard is that face-to-face, embodied interaction provides a context too rich to ever be encompassed in a "virtual," technologically mediated, world. Since the relationship between an educator and his or her student falls within the variety of human relationships, it demands that we ask if and when this relationship reaches beyond the framework adequately captured through technological mediation. Such is our goal here: to sound a cautionary note that encourages this type of questioning.

Anyone who has tried to maintain a long-distance relationship has experienced the dissatisfaction of being confined to letters, phone calls and email. Nonetheless, these technologically mediated communications are often better than nothing at all.[27] The situation with "virtual classrooms" may be similar. It seems likely that Internet based education is, within certain boundaries, better than no education at all.[28] However, there are good reasons for believing that unqualified optimism about "online environments" is misplaced. Unless Internet based education can be clearly demonstrated to be *better* than real-classroom interaction, we should be concerned about the apparent eagerness to move more and more students into virtual classrooms at the cost of diminishing more traditional classroom environments.[29] The question to what degree we should willingly advocate using the Internet for education and "bringing people together" requires thoughtful examination. At the very least then, we conclude that such examination requires careful sorting of content appropriately suited to computer mediated instruction from content not amenable to quality education without there being actual persons, located in actual classrooms, conducting face-to-face discussion and dialogue.

ENDNOTES

[1] Todd Oppenheimer, "The Computer Illusion," *The Atlantic Monthly* (July, 1997), on the Internet at www.TheAtlantic.com/issues/97jil/computer.htm p. 4.

[2] The expression "virtual classroom" seems to have been originated, in print, by Starr Roxanne Hiltz in her *The Virtual Classroom: Learning Without Limits via Computer Networks* (Norwood, NJ: Ablex Publishing Co., 1994).

[3] Gary Goettling, "Going the Distance," *Georgia Tech Alumni Magazine* (Spring, 1999), p. 30.

[4] "Computers in Distance Education: Guide #7," (College of Engineering, University of Utah: October 1995), p. 2. Similarly, Richard A. Crofts, Commissioner of the Montana University System, quoted by Albert Borgmann in *Holding On to Reality: The Nature of Information at the Turn of the Millennium* (Chicago: The University of Chicago press, 1999), pp. 259 – 260n34, has said that "[F]or the first time in the history of American higher education, information technology provides the opportunity to increase access to higher education, improve the quality of students' learning experiences, enhance the faculty role as teacher/scholar/learner, and control the costs of education – simultaneously."

[5] Robert Kraut, Vicki Lundmark, Michael Patterson, Sara Kiesler, Tridas Mukopadhyay, and William Scherlis, "Internet Paradox: *A Social Technology That Reduces Social Involvement and Psychological Well-Being?*," (September, 1988) on the Internet at www.apa.org/journals/amp/amp5391017.html, p. 11. Originally published in *American Psychologist*, v. 53, n.9 (September, 1998), pp. 1017 – 1031.

[6] *ibid.* Also see Michelle M. Weil and Larry Rosen, "Commentary on the HomeNet Study," on the Internet at http://technostress.com/homenet.htm.

[7] Fred Moody, "Commentary on Technology: A Modest Experiment," (September 4, 1998), on the Internet at http://archive.abcnews.go.com/sections/tech/FredMoody/moody980904.html. David Ignatius, in his "Grinch may lurk on mountaintop of online products," *Atlanta Journal-Constitution* (December 12, 1999), p. G1, discusses this concern with respect to the growing use of the Internet as a "cyber-replacement" for the traditional shopping experience.

[8] For two wider discussions of the effect of computer mediated communication on human interactions, see Joseph B. Walther and Judee K. Burgoon, "Relational Communication in Computer-Mediated Interaction," *Human Communication Research*, v. 19, n. 1 (September, 1992), pp. 50 – 88, and Russell Spears and Martin Lea, "Panacea or Panopticon? The Hidden Power in Computer-Mediated Communication," *Communications Research*, v. 21, n. 4 (August, 1994), pp. 427 – 459.

[9] See Hubert Dreyfus, "Anonymity versus commitment: The dangers of education on the Internet," *Ethics and Information Technology*, v. 1, n. 1 (1999), pp. 15 – 21. Also, his earlier "Dangers and Vistas on the Information Highway: The Future of Information Technology as Seen in 1850 by Søren Kierkegaard" (presented at the Stanford Humanities Center, March 1997 [unpublished]).

[10] Hubert Dreyfus, "Dangers and Vistas on the Information Highway: The Future of Information Technology as Seen in 1850 by Søren Kierkegaard" (presented at the Stanford Humanities Center, March 1997 [unpublished]) p.3.

[11] *Søren Kierkegaard's Journals and Papers,* 7 vols., tr. & ed. Hong & Hong (Bloomington: Indiana University Press, 1967), entry #4233.

[12] *ibid.*, entry #2170

[13] *ibid.*, entry #2152

[14] *ibid.*, entry #2150

[15] Sherry Turkle, *Life on the Screen: Identity in the Age of the Internet* (NY: Simon and Schuster, 1995), p. 207.

[16] Amy Harmon, "Researchers Find Sad, Lonely World in Cyberspace," on the Internet at www.nytimes.com/library/tech/98/08/biztech/articles/30depression.html. Don Tapscott, in his *Growing Up Digital: The Rise of the Net Generation* (New York: McGraw-Hall, 1998), makes an even stronger claim. He writes, p. 107, that:

> …Digital kids are learning precisely the social skills which will be required for effective interaction in the digital economy. They are learning about peer relationships, about teamwork, about being critical, about how to have fun online, about friendships across geographies, about standing up for what they think, and about how to effectively communicate their ideas.

[17] For discussions of how such anonymity may have a number of positive benefits, see R.F. baumeister, "A self-presentational view of social phenomena," *Psychological Bulletin*, v. 91 (1982), pp. 3 – 26. Also see V. Dubrovsky, S. Kiesler, and B. Sethna, "The equalization phenomenon: Status effects in computer-mediated and face-to-face decision making groups," *Human Computer Interaction*, v. 6 (1991), pp. 119 – 146.

[18] Peter Danielson, "Pseudonyms, Mailbots, and Virtual Letterheads: The Evolution of Computer-Mediated Ethics," in *Philosophical Perspectives on Computer-Mediated Communication,* edited by Charles Ess (NY: State University of New York Press, 1996), p. 74.

[19] See Robert S. Fortner, "Excommunication in the Information Society," *Critical Studies in Mass Communication*, v. 12 (1995), pp. 134 – 136, 144ff.

[20] Ron Barnette, "Teaching Philosophy in Cyberspace," in *The Digital Phoenix: How Computers are Changing Philosophy*, edited by Terrell Ward Bynum and James H. Moor (Oxford: Blackwell Publishers Ltd., 1998), p. 324.

[21] *ibid.*

[22] *ibid.,* p. 325. It is important to note that Barnette sees this as a virtue of education mediated through the Internet by the creation of virtual classrooms. We have a much different take on Barnette's characterization.

[23] Or, as the protagonist of Turkle's story puts it, the feeling that somehow "Real life gave me too much information."

[24] See Kenneth Gergen, *The Saturated Self: Dilemmas of Identity in Contemporary Life* (NY: Basic Books, 1991), p. 17.

[25] Gary Shank and Donald Cunningham, "Mediated Phosphor Dots: Toward a Post-Cartesian Model of Computer-Mediated Communication via the Semiotic Superhighway," in *Philosophical Perspectives on Computer-Mediated Communication,* edited by Charles Ess (NY: State University of New York Press, 1996), p. 39.

[26] Postman's view as characterized by Andrew Leonard in his "Internet U.: A New Documentary Looks at the Benefits and Hazards of the Net as Global Lecture Hall," on the Internet at www.salon.com/21st/reviews/1998/09/04review.html

[27] See, for example, Molly Masland, "Net gives seniors new worlds to visit," on the Internet at www.msnbc.com/news/178302.asp who discusses how the Internet provides a valuable tool for communication by seniors who might not otherwise have any easy way to regularly communicate with other people.

[28] See, for example, Mike Brunker, "Building the 'Great Wired North'," on the Internet at www.msnbc.com/news/210223.asp for an example of how the use of the Internet, as a supplemental tool in the educational process, holds the promise of enhancing students' education.

[29] See Eugene Heath, "Two Cheers and a Pint of Worry: An On-Line Course in Political and Social Philosophy," *Teaching Philosophy,* v. 20, n. 3 (September, 1997), pp. 277 – 300. Also see *op. cit.,* Borgmann, pp. 207 – 208, and *op. cit.,* Oppenheimer.

Chapter 16

Manufacturing Social Responsibility Benchmarks in the Competitive Intelligence Age

James Douglas Orton
University of Nevada, Las Vegas, USA

I am in the competitive intelligence version of a witness protection program. After six years as a spy to the French eye and as a traitor to the American eye, I am becoming comfortable blending back in to the relatively monocultural population of unquestioning Americans. No accents, no suspicions, no guarded words, no misinformation, no handlers, no fauxpas, no culture shock–I have come in from the competitive intelligence cold. Although I can feel myself being reabsorbed into the warm American Emersonian oversoul, I am haunted by the guilt of the double agent. What damage did I do in the last six years?

The Americans believe I sold out the secrets of the American-dominated Fortune 500 for 30 pieces of Parisian silver. Why, they wonder, would a young business school professor (trained in the American heartlands of Utah, Texas, and Michigan) abandon the US economy to go work for another country? When I showed French MBAs, through Harvard Business School cases, that Harley Davidson, Corning Glass, and Caterpillar Tractor had weaknesses that could be skillfully exploited from outside the US, wasn't I being a traitor to my country's economy? When I taught French doctoral students the arcane arts of publishing articles in American business academic journals, wasn't I taking journal space away from American doctoral students? When I taught French executives how Americans built corporate strategies from thousands of small wins, wasn't I aiding and abetting the enemies of America's economic security? Finally, though, when I agreed to teach an elective course to my elite international students on strategic intelligence, didn't I commit the unforgiveable sin of raising up a generation of spies who might torment my country for years to come?

The French believe I was always under the control of CIA handlers. When, every year, I would ask for a raise, the response from senior colleagues would be that I didn't need one, since I already had two salaries: one from the CIA and one from the French business school. The French foreign minister started referring to the United States as a hyperpower. Graffiti started springing up around Paris saying "America dehors l'Europe" (Get America out of Europe). In that context, when I was caught in their libraries, studying their internship reports, and using their Reuters subscription, it was clear to the French that I was an economic intelligence agent for the US government tunneling for information on Totale, Danone, Schneider, Aerospatiale, Air France and Carrefour.

One case study that captures my six years as a competitive intelligence agent involved a meeting of French competitive intelligence officers from 40 large corporations. My French business school employers dipped into their training budget to send me to a seminar in downtown Paris on economic intelligence. For two days, 60 of us listened to 10 presenters explain the state of the art of economic intelligence. A pharmaceutical firm told us how they had transformed their sales network into a business intelligence system by developing electronic contact reports (Thietart & Vivas, 1981). An aerospace company explained how their small intelligence unit was using Web-clipping software to send information to relevant sections of their organization (Gibbons & Prescott, 1996). A big-picture thinker formerly with L'Oreal used Rene Magritte paintings to expand our minds toward the creative use of information. Hubert Lesca presented academic research he has conducted with his doctoral students at the University of Grenoble on removing blockages in the intelligence cycle. Yves-Michel Marti described the elaborate intelligence cycle his consulting firm uses to generate economic intelligence for their clients (Martinet & Marti, 1995). Bernard Besson explained, from his police background, how to conduct counterintelligence operations (Besson & Possin, 1997). Frederic Jakobiak knit the presentations together as a host and commentator.

Early in the two-day conference, one of the presentations came from a part-time instructor at Marne-la-Vallee. The professor explained that a large part of the operation at Marne-la-Vallee involved Internet surfing on the Web sites of American multinationals. They found, though, that some of the American sites were location sensitive, so that a search conducted from an American address would yield different screens than a search conducted from a French address. Furthermore, the young professor was horrified to discover that some of the American multinational Web sites used "sniffers" and "cookies" to try to identify the location and identity of the remote economic intelligence surfers at Marne-la-Vallee. He explained how to create a buffer or indentities by searching through a chain of addresses. He implied that the National Security Agency, the CIA, and IBM were working together to plant viruses, false data, and identification flags on the computers of French economic intelligence agents, and that the French would fight back through viruses of their own, in a spirit of "Cocorico," a French word associated with their national symbol of the rooster, implying "we got you."

In this climate, I found myself at the traditional business meal of eight people sitting around a table eating a salad. All seven of my new colleagues were competitive intelligence officers at French multinationals. After we all introduced ourselves to each other, the woman two seats to my left asked why the National Security Agency had interfered with a contract between the Brazilian government and the French defense company Thomson Electronics. She worked for Thomson Electronics and felt that the US government had interfered in the negotiations by passing on NSA-procured eavesdropping data to the Brazilian government to steer the contract toward the American firm Raytheon. I told the woman that the story I had heard was that the NSA picked up a bribe offer to the Brazilian defense minister and communicated that information to the Brazilian President, who then overruled the Thomson contract in favor of the Raytheon contract. There was silence around the table for a few awkward moments.

The themes explored elsewhere in this volume, on the intersection between information technology and social responsibility, take on new shapes when considered in the context of competitive intelligence. Using the (probably) apocryphal Thomson-Raytheon story as a launching point, this chapter will explore the emergence of social responsibility benchmarks in the competitive intelligence age. This analysis is heavily flavored by my own experiences trying to understand the French approach to competitive intelligence. The paper reviews attempts by competitive intelligence agents in the US and France to manufacture Social Responsibility benchmarks in the contexts of covert operations, competitive strategy, corporate intelligence, economic security, economic intelligence, and economic warfare. The conclusion of the paper will argue that the construction of social responsibility is a local-level human accomplishment, not a global-level rational standard. Furthermore, the paper implies that the burden of social responsibility lies more heavily on the successful economic oppressor than the unsuccessful economic resistance.

COVERT OPERATIONS

Loch Johnson, a political science professor at the University of Georgia, proposed a "partial ladder of escalation for intelligence operations," in reverse order, from "extreme options" through "high-risk options" through "modest intrusions" to "routine operations" (Johnson, 1996, p. 147).

Threshold Four: Extreme Options

38. Use of chemical-biological and other deadly agents
37. Major secret wars
36. Assassination plots
35. Small-scale coups d'etat
34. Major economic dislocations: crop, livestock destruction
33. Environmental alterations
32. Pinpointed retaliation against noncombatants
31. Torture
30. Hostage taking
29. Major hostage-rescue attempts
28. Theft of sophisticated weapons or arms-making materials
27. Sophisticated arms supplies

Threshold Three: High-Risk Options

26. Massive increases of funding in democracies
25. Small-scale hostage rescue attempt
24. Training of foreign military forces for war
23. Limited arms supplies for offensive purposes
22. Limited arms supplies for balancing purposes
21. Economic disruption without loss of life
20. Large increases of funding in democracies
19. Massive increases of funding in autocracies
18. Large increases of funding in autocracies
17. Sharing of sensitive intelligence
16. Embassy break-ins
15. High-level, intrusive political surveillance
14. High-level recruitment and penetrations
13. Disinformation against democratic regimes
12. Disinformation against autocratic regimes
11. Truthful but contentious information in democracies
10. Truthful but contentious information in autocracies

Threshold Two: Modest Intrusions

9. Low-level funding of friendly groups
8. Truthful, benign information in democracies
7. Truthful, benign information in autocracies
6. Stand-off TECHINT against target nation
5. "Away" targeting of foreign intelligence officer
4. "Away" targeting of other personnel

Threshold One: Routine Operations

3. Sharing of low-level intelligence
2. Ordinary embassy-based observing and conversing
1. Passive security measures; protection of allied leaders

In general, this is a list of options that have been used by governments, not corporations. However, there have been reported cases of corporations attempting (25) small-scale hostage rescue attempts (EDS in Iran), (18) large increases of funding in autocracies (ITT in Chile), and (17) sharing of sensitive intelligence (German engineering firms in Iraq). The list of covert operations provides a starting point for our study of the manufacturing of ethics in an age of competitive intelligence. How far up this list will corporations go as they seek to understand and influence their environments? Presumably corporations are more tightly constrained in their actions than governments, and we will see an expansion of types of actions in the first threshold, such as misrepresentation of facts, bribery of competitors' employees, and theft of information.

COMPETITIVE STRATEGY

An understanding of competitive intelligence requires an understanding of the history of competitive strategy. The era of firms lasted from 1500-1865. Firms are small, family-owned, entrepreneur-led businesses with 20 or fewer employees: family farms, small mills, bakeries, grocery stores, shoe cobblers, and blacksmith shops. Firms compete in large, "pure" markets composed of similar firms. The era of bureaucracies lasted from 1865-1944. Bureaucracies were composed of large conglomerations of firms: e.g., General Mills, General Foods, and General Motors. The railroad, telegraph, automobile, and telephone made it economically feasible to coordinate large bureaucracies over long distances. The era of networks lasted from 1944-2001. Networks relied on improved mobility and information technology to create alliances among numerous firms in loosely coupled, organic, international networks composed of autonomous, empowered, intelligent actors. It was not until the era of networks that business operations became complex enough to require the emergence of "competitive strategies."

Herbert Simon and his colleagues shifted the attention away from bureaucratic structures toward network strategies in the 1950s (Simon, 1947, 1955, 1957a, 1957b, 1976). They laid the foundation for later discussions of organizations as organized anarchies (Cohen, March, & Olsen, 1972; March & Olsen, 1976) and loosely coupled systems (Weick, 1976). The case study that made Simon and his colleagues' theories tangible to researchers was the Cuban Missile Crisis, deftly analyzed by Allison through three lenses: rational actor, bureaucratic politics, and organizational processes. During a crisis, a firm moves from the chaotic organizational process model, to the factional Bureaucratic Politics model, to the ordered rational actor model (Allison, 1969, 1971; Allison & Zelikow, 1999). At about the same time as the Cuban Missile Crisis, Kennedy's Harvard rowing team colleague Alfred D. Chandler published an account of bureaucratic changes in the 1920s in a book he labeled *Strategy and Structure*. Chandler used the military strategy metaphor from his teaching at the Naval War College to make sense of the business structural changes he studied in General Motors, DuPont Chemicals, Sears & Roebuck, and Standard Oil (Chandler, 1962). Harvard Business School combined the Allison case with the Chandler cases to create the field of business strategy.

In 1980, Michael Porter crystallized an industrial/organizational economics view of business strategy in his book *Competitive Strategy*. Generations of business students have memorized the five forces model in Chapter 1 of the book, but few students or their professors remember that the five forces model is data-hungry and requires a great deal of competitive intelligence research, described in Chapter 3 and Appendix B:

> Answering these questions about competitors creates enormous needs for data. Intelligence data on competitors can come from many sources: reports filed publicly, speeches by a competitor's management to security analysts, the business press, the sales force, a firm's customers or suppliers that are common to competitors, inspection of a competitor's products, estimates by the firm's engineering staff, knowledge gleaned from managers or other personnel who have left the competitor's employment, and so on (Porter, 1980).

The public, popular side of Porter's work has been the strategy content analyses, but the hidden, dark side of Porter's work is competitive intelligence collection and analysis.

CORPORATE INTELLIGENCE

After Porter, researchers at other schools started to focus on how corporations gather political, technological, cultural, and "violence" intelligence (Eells & Nehemkis, 1984). Columbia professor Richard Eells and UCLA professor Peter Nehemkis described large business organizations as "private governments" and asserted that these "polycorporations" needed intelligence units:

> Without question, in our judgment, the large multinational (or polycorporation, as we call it) should establish well-conceived, effective, professionally-staffed intelligence units within its management structures. It goes without saying that close and effective supervision by the chief executive officer of [the] company's intelligence unit is a sine qua non for its effective and useful operations–operations that are beneficial to the the corporation's own policies and the chief executive's own decision-making (p. 221).

Eells and Nehemkis found that studies of corporate intelligence quickly lead to questions of social responsibility: "As the research for this book came to an end it became clear that further studies should be undertaken, specifically of the implications for public policy of the private intelligence community's growth, especially in the matters of privacy, morality and ethics" (p. xii). Eells and Nehemkis tried to patch up this missing discussion in their book by listing 12 questions from Professor R. R. Nash (1981) about ethical business decisions:

1. Have you defined the problem accurately?
2. How would you define the problem if you stood on the other side of the fence?
3. How did this situation occur in the first place?
4. To whom and to what do you give your loyalty as a member of the corporation?
5. What is your intention in making this decision?
6. How does this intention compare with the probable results?
7. Whom could your decision or action injure?
8. Can you discuss the problem with the affected parties before you make your decision?
9. Are you confident that your position will be as valid over a long period of time as it seems now?
10. Could you disclose without qualm your decision or action to your boss, your CEO, the board of directors, your family, society as a whole?
11. What is the symbolic potential of your action if understood? If misunderstood?
12. Under what conditions would you allow exceptions to your stand?

If our list of covert operations is too focused on governments in general, this list is too focused on business decisions in general.

ECONOMIC SECURITY

Depending upon party affiliations, George Bush and Bill Clinton share credit for launching the boom in international economic intelligence. Johnson gives a nod to Bush: "A 1991 review of intelligence priorities, initiated by President Bush, led to a dramatic allocation of resources away from old Cold War concerns toward new economic targets, as the world marketplace became an ever more important battlefield for America" (Johnson, 1996, p. 147). As part of those discussions, retired Director of Central Intelligence Admiral Stansfield Turner launched a provocative proposal in foreign affairs in fall of 1991:

> The preeminent threat to US national security now lies in the economic sphere. ...We must, then, redefine "national security" by assigning economic strength greater prominence. ...If economic strength should now be recognized as a vital component of national security, parallel with military power, why should America be concerned about stealing and employing economic secrets (Turner, 1991).

The larger share of the credit, though, according to Johnson (1996), should go to Clinton:

> The question of economic competitiveness served as a centerpiece in the 1992 presidential campaign of Bill Clinton. On the eve of assuming the presidency, he vowed to "make the economic security of our nation a primary goal of our foreign policy." Coupled with his abiding attention to domestic economic issues, foreign economic policy became one of the president's foremost concerns during the first years of his administration. Early in office he created a National Economic Council (NEC), touted as equal in status to the National Security Council. Long a maid in waiting to defense issues, matters of trade and aid had risen high on the national security agenda. "It's the economy, stupid!' had been the mantra among Clinton's campaign strategists in 1992; now, within the government's community of national security planners, the slogan seemed to be, "It's the economy, stupid!" (pp. 146-147)

The collective wisdom of intelligence community think tanks quickly chimed in with a collective opinion: The involvement of the US government in microeconomic intelligence was a really bad idea. In a paper presented on April 8, 1993, Randall M. Fort listed 20 reasons why it would be a bad idea. (1) The "economic threats" have always existed, are often caused by ourselves, and are probably better labeled "economic challenges." (2) Economic benefits are not black and white, but intertwined between countries, such as in the case of a Honda plant in the United States providing jobs for Americans; this idea is coded after the title of a Robert Reich article, "Who Is 'Us'?" (3) How would we decide which industries and which firms received economic intelligence? (4) Competition for economic intelligence would become a subsidy allocated on the basis of "political clout." (5) American firms are involved in alliances with non-American firms. (6) It would be difficult to protect sources and methods. (7) Productivity of intelligence assets might decline, be harmed, or dry up if the sources feel their data is being used for economic purposes. (8) Recipients of economic intelligence could leave a US firm to work for a foreign firm. (9) Retooling to provide intelligence to the private sector would be expensive. (10) Uncertain information could lead to conflicts between the US firms and the US government. (11) Economic competition with foreign governments would reduce the U.S.'s capacity to create military and diplomatic alliances. (12) US firms do not want to be associated with the US intelligence community in suppliers' and consumers' minds. (13) US firms do not consider the US intelligence community a reliable source of information. (14) Supplying intelligence to US firms would require legal changes in the enabling statutes and executive orders, in the Trade Secrets Act, in the wire fraud statutes, and in the Communication Act and Foreign Intelligence Surveillance Act. (15) Supplying economic intelligence to firms could subject the US government to civil litigation by harmed constituents. (16) Just because the French, the Israelis, the Germans, the Japanese, and the South Koreans are involved in nationally sponsored economic intelligence, that does not make it morally acceptable. (17) US sponsorship of economic intelligence would unleash more dangers for US businesspeople abroad. (18) US economic intelligence might invite economic intelligence from nations that are better at it than we are. (19) Intelligence agencies would be exposed to corruption from US firms who might benefit from preferential intelligence analyses. (20) Finally, in one of the most frequently cited arguments, US intelligence officers would not be motivated by the economic intelligence task: "Some years ago, one of our clandestine officers overseas said to me: 'You know, I'm prepared to give my life for my country, but not for a company.' That case officer was absolutely right" (Fort, p. 196, citing Robert M. Gates, Speech to the Economic Club of Detroit, April 13, 1992).

Fort summarized his arguments against US government involvement in economic intelligence by tracking the differences between DCI James Woolsey's statements in his confirmation hearings with his statements one year later:

> Then-DCI James Woolsey raised eyebrows and expectations during his Senate confirmation hearing when he described economic espionage as "the hottest current topic in intelligence policy." Subsequent news stories indicate that he has reached some unenthusiastic conclusions about such an effort. One year later, he was quoted as stating that such a program would be "fraught with legal and foreign policy difficulties." Woolsey's disapproving tone is not surprising. Anyone who gets past the rhetoric about economic competitiveness and closely examines the nuts and bolts

of how an economic espionage program is supposed to work cannot fail to reach the same negative conclusions (p. 196)

Despite Fort's explanation of why US microeconomic intelligence is a bad idea, the genie was already out of the bottle in 1992. The signal that the US sent was that competitive microeconomic intelligence was on the agenda. That shot heard round the world in 1992 gave other governments the motivation to ramp up their own competitive intelligence activities.

ECONOMIC INTELLIGENCE

The French, for example, asked Prefet Henri Martre to preside over a 1994 study titled "Intelligence economique et strategie des entreprises" (Economic Intelligence and Business Strategy). The report identified five trends: (1) Economic intelligence is increasingly recognized internationally as a criterion for competitiveness; (2) Economic intelligence is the raw material for the creation of a new industry of competitive intelligence firms; (3) Economic intelligence should be retained by the organization as a form of knowledge capital; (4) Economic intelligence needs to be supported by governmental units at the levels of regions, countries, and territories; (5) Economic intelligence is being defined by the Americans and others as a national security issue (Commissariat General du Plan Documentation Francaise, 1994). To respond to these five trends, the Martre report encouraged French firms, French regions, and the French state to improve their economic intelligence capabilities.

The Martre report argued that firms such as Exxon, General Electric and Boeing had used economic intelligence units since as early as 1972. In addition, the report asserted that generalized consulting firms such as McKinsey have economic intelligence capabilities, and that specialized consulting firms staffed by former intelligence officers are emerging to create a larger system of firms working together for the benefit of the US economy (p. 63).

The Martre report also argued that national cultures provided significant backdrops for the creation of intelligence systems:

It is not British Petroleum's economic intelligence tools that create their excellence in this domain, but its culture and history, which are intimately linked to the intelligence culture that the British Empire developed during its history. Economic intelligence systems created in China, Japan, the Middle East, the United States, Great Britain and Germany all have cultural roots [my translation] (pp. 64-65).

The Martre report also argued that nations created and supported economic intelligence units at two levels, defensive and offensive:

It is important to distinguish two levels of analysis. The first level is the preservation of employment and national sovereignty, and no industrial country hides the fact that it is operating this type of economic counterintelligence or industrial counterintelligence, when these types of economic intelligence attacks pass legal norms. The second level of analysis is the protection of a threatened industry. In this context, the encouragement of exports and the maintenance of economic competitiveness is as important as the protection of the national inheritance. Because of "national conscience" and "economic patriotism," economic intelligence and industrial intelligence become part of the domain of activities handled by countries' Industry Ministries, External Commerce Ministries, and economic institutes [my translation] (p. 68).

The Martre report became the foundation in France for an enthusiastic project of state-sponsored competitive intelligence programs. Admiral Lacoste (of Rainbow Warrior fame) started a DESS d'ingenierie de l'intelligence economique (a master's program in the engineering of economic intelligence) at Université de Marne-la-Vallee, east of Paris. The Université de Poitiers launched a DESS d'information et culture stratégique (a master's program in strategic information and culture). The Universite d'Aix-Marseille launched a DEA de veille technologique (a pre-doctoral degree in the surveillance of emerging technologies). The CERAM-ESC Nice business school launched a specialized master's degree in economic intelligence. A variety of regional economic intelligence initiatives were created throughout France to sensitize French executives to the importance of creating a national competitive intelligence culture.

COMPETITIVE INTELLIGENCE

Meanwhile, back in the United States, a new career path was developing around the topic of competitive intelligence. One of the best statements of the developing field is Larry Kahaner's 1996 book, *Competitive Intelligence: How to Gather, Analyze, and Use Information to Move Your Business to the Top.* Kahaner's case studies (which are rare in competitive intelligence studies) included Avon and the Marriott Corporation. Avon hired a consulting company to analyze the garbage left in trash bins on public property by Mary Kay Cosmetics, in order to find out what Mary Kay Cosmetics' strategic plans were. The collection method was legal, but Kahaner asks the question, "Yes, but is it ethical?" The two firms came to the conclusion that Avon could reassemble the shredded garbage as long as a Mary Kay Cosmetics employee was there to see what they reassembled. In the Marriott case, Marriott hired an executive recruiting firm to interview executives in the economy hotel business, which the search firm then fed back to Marriott to help them craft their strategy for entering this new market. Although Marriott did hire some of the executives interviewed, it seems that the primary rationale for the operation was intelligence collection, rather than hiring. Here again, Kahaner asks about the appearance of the operation as ethical or unethical.

Another book that helped shape the development of competitive intelligence as a career path is Leonard M. Fuld's *The New Competitor Intelligence: The Complete Resource for Finding, Analyzing, and Using Information About Your Competitors*. From years of experience trying to conduct competitive intelligence, Fuld & Company (1995) created "the ten commandments of legal and ethical intelligence gathering":

(1) Thou shalt not lie when representing thyself.
(2) Thou shalt observe thy company's legal guidelines as set forth by the Legal Department.
(3) Thou shalt not tape-record a conversation.
(4) Thou shalt not bribe.
(5) Thou shalt not plant eavesdropping devices.
(6) Thou shalt not deliberately mislead anyone in an interview.
(7) Thou shalt neither obtain from nor give to thy competitor any price information.
(8) Thou shalt not swap misinformation.
(9) Thou shalt not steal a trade secret (or steal employees away in hopes of learning a trade secret).
(10) Thou shalt not knowingly press someone for information if it may jeopardize that person's job or reputation (Fuld, 1995).

These recommendations provide a more helpful set of beginning benchmarks for the construction of social responsibility than our two previous lists, Johnson's ladder of intrusion in covert operations and Nash's questions for general business decisions.

In a preemptive move intended to protect their industry against government intervention, the Society of Competitive Intelligence Professionals (often estimated to have 5,000 members) drafted their own code of ethics:

(1) To continually strive to increase respect and recognition for the profession on local, state and national levels.
(2) To pursue his or her duties with zeal and diligence while maintaining the highest degree of professionalism and avoiding all unethical practices.
(3) To faithfully adhere to and abide by his or her company's policies, objectives, and guidelines.
(4) To comply with all applicable laws.
(5) To accurately disclose all relevant information, including the identity of the professional and his or her organization, prior to all interviews.
(6) To fully respect all requests for confidentiality of information.
(7) To promote and encourage full compliance with these ethical standards within his or her company, with third-party contractors, and within the entire profession.

This preemptive move by the competitive intelligence industry did not hold, and President Clinton signed The Economic Espionage Act on October 11, 1996. The act defines trade secrets broadly to include "all forms and types of information including financial, business, scientific, technical, economic, and engineering. It includes plans, formulas, designs, prototypes, methods, techniques, processes, procedures, computer codes, and so on" (p. 244). Penalties for stealing a trade secret include up to 10 years in prison, individual-level fines of up to $500,000, and corporate-level fines of up to $5 million. If the theft benefits a foreign entity, the penalties can rise to 15 years imprisonment and a corporate-level fine of up to $10 million.

The benchmarks that the competitive intelligence industry is generating for itself–Kahaner's case studies, Fuld & Company's ten commandments, SCIP's guidelines, and the Economic Espionage Act's sanctions–are helpful within the US context. More work needs to be done, though, on competitive intelligence activities between countries.

ECONOMIC WARFARE

One of my favorite colleagues, Patrick Lemattre, cursed with an impish sense of humor, invited me to attend one of his courses. He was hosting Christian Harbulot, who had had a hand in the Martre report, then become operational director at French military intelligence economic spin-offs DCI and Intelco, and had become the "tête pensant" or guru of the Ecole de Guerre Economique, or the School of Economic Warfare. The school opened in October 1997 under the auspices of the Ecole Superieure Libre des Sciences Commerciales Appliquees, under the leadership of General Jean Pichot-Duclos (Merchet, 1997).

Due to another course responsibility, I was not able to introduce myself to Christian Harbulot before the presentation began, so he assumed he was speaking to a room of French students and faculty members. He had transparencies showing that the Americans controlled the United Nations, the Organization for Economic Cooperation and Development, the Church of the Reverend Moon, and the Republic of Germany. He said that the first assault on France would be through Coca-Cola, McDonalds, and Disney, all of which are designed to weaken the attachment of citizens to their national cultures. He said that the second assault would be through music, movies, and television, which would be used to propagate American values around the world. The third assault, he explained, would come from General Motors, IBM, Hewlett-Packard, Microsoft, and other American multinationals, which would then subjugate the world populations as captive employees.

How then, can France defend itself against this organized, controlled, centrally planned offensive? The explicit script was to boycott American intrusions into France: "As a Frenchman," Harbulot said, "I no longer drink Coca-Cola." The implicit script was more subtle and can be summarized in the phrase, "Tous les coups sont permis contre les americains" or "Against Americans, there are no rules–anything goes." If France and other cultures around the world feel that they are under an organized attack from an American economy, discussions of fairness, ethics, and social responsibility become irrelevant. Groupthink and a siege mentality set in (Janis, 1972), and Americans are painted as the immoral aggressors, while the French resistance can only be painted as moral and heroic.

CONCLUSION: MANUFACTURING SOCIAL RESPONSIBILITY BENCHMARKS

All of this discussion of covert operations, competitive strategy, corporate intelligence, economic security, economic intelligence, competitive intelligence, and economic warfare brings us back to the case study of the NSA eavesdropping on Thomson Electronics to shift a Brazilian defense contract away from Thomson Electronics toward Raytheon. Where does social responsibility lie in this case?

First, here's what I said at the lunch table. I said that the coordination between the economic sector and the governmental sector in the US was not at all as tight as it was in France, and my French colleagues were projecting coordinated action onto a situation where there was none. Instead, each firm was acting in its own economic self-interest, and the French were misinterpreting the cumulative effect of profit-seeking at the firm level as a grand conspiracy at the national level. I said that most Americans didn't even know where France was, so why would they spend all their time trying to overrun it economically? I argued that French firms should respond to American firms, not engineer a French societal response to a presumedly coordinated American societal attack. I also argued that it seemed that French consultants were fueling an intelligence arms race by misrepresenting the American intelligence threat. Each protest that there was no conspiracy led to new questions about how the conspiracy was structured until, thankfully, other topics emerged for discussion.

Now, though, I would not have been so eager to defend the American point of view. Instead, I see both sides of the conflict as flawed, self-serving ethical frameworks.

To the French, the Americans have moved from being an ally in a conflict between two blocs, the Soviets and the Americans, to an often-destructive singular "hyperpower." The French–as they did with the Germans in World War II–have constructed, enacted, and manufactured an ethical code that allows them enormous flexibility in defending themselves against an invasion by a hyperpower. To use a variety of means to steer the contract toward Thomson Electronics and away from Raytheon, including the time-honored tradition of offering a secret "pot de vin," is fair, especially given the enormous advantages of the American cultural barbarians.

To the Americans, the French are just another market to conquer in a firm-by-firm quest for increased market share, increased sales, and increased profits. The stubborn French resistance frustrates the Americans, who complain that the French are exploiting a home-field advantage to skew the outcome of the competition. To use National Security Agency assets to correct an unfair influence from France, and to steer the contract toward Raytheon, is–to American eyes–a justifiable use of national intelligence resources.

How do we solve such an intractable case? We know from research on sense-making processes that human beings interpret ongoing streams of events in ways that reinforce the significance of their own identities, through intensive social discussions with people they work with, and through flawed, outdated, retrospective models (Weick, 1995). Is there any reason to believe that ethics-making is any more rational, objective, or precise than sense-making?

So, no matter how many times we attempt to rule out the chaos of the ethics of competitive intelligence, it will always come down to whether or not individuals can construct an ethical framework that allow themselves to feel good when they look in the mirror. And, in an example of how easy it is for people to feel good when they look in the mirror, 75% of American males think they are in the top 25% of athletic ability. French friends of Thomson look good in their mirrors, and American friends of Raytheon look good in their mirrors, but they are both flawed–one bribes and one eavesdrops, and the battle is only over which is the least unethical course of action.

My conclusion to the case is thus rather skeptical. Although competitive intelligence professionals are well on the road to constructing social responsibility benchmarks within homogenous cultures, the larger problem of international social responsibility benchmarks is going to take a great deal of work, given the enormous forces that encourage local-level construction of benchmarks, rather than global rational benchmarks. I suspect that the greater burden should lie with the economic "aggressors" who are having the most success, rather than the companies and countries constructing an economic "resistance." New cross-national case studies should be collected to help find the appropriate balance between nations and corporations in this question for social responsibility benchmarks in the age of competitive intelligence.

ACKNOWLEDGMENTS

This paper was financed by grants from the HEC Foundation (1994-2000) and the UNLV New Investigator Award (2000-2002).

REFERENCES

Allison, G. T. (1969). Conceptual models and the Cuban Missile Crisis. *The American Political Science Review*, 63, 689-718.

Allison, G. T. (1971). *Essence of Decision: Explaining the Cuban Missile Crisis*. Boston, MA: Little, Brown, and Co.

Allison, G. and Zelikow, P. (1999). *Essence of Decision: Explaining the Cuban Missile Crisis*. (Second ed.). New York: Addison Wesley Longman.

Besson, B. and Possin, J. C. (1997). *Du renseignement a l'intelligence strategique: Detecter les menaces et les opportunites pour l'entreprise*. Paris: Dunod.

Chandler, A. D., Jr. (1962). *Strategy and Structure: Chapters in the History of the Industrial Empire*. Cambridge, MA: MIT Press.

Cohen, M. D., March, J. G. and Olsen, J. P. (1972). A garbage can model of organizational choice. *Administrative Science Quarterly*, 17, 1-25.

Commissariat General du Plan Documentation Francaise, L. (1994). *Intelligence Economique et Strategie des Eentreprises*.

Eells, R. and Nehemkis, P. (1984). *Corporate Intelligence and Espionage: A Blueprint for Executive Decision Making*. New York: Macmillan.

Fuld, L. M. (1995). *The New Competitor Intelligence: The Complete Resource for Finding, Analyzing and Using Information About Your Competitors*. New York: John Wiley & Sons, Inc.

Gibbons, P. T. and Prescott, J. E. (1996). Parallel competitive intelligence processes in organizations. *International Journal of Technology Management*, 11(1-2), 162-178.

Janis, I. L. (1972). *Victims of Groupthink: A Psychological Study of Foreign-Policy Decisions and Fiascoes*. Boston, MA: Houghton, Mifflin.

Johnson, L. K. (1996). *Secret agencies: US Intelligence in a Hostile World*. New Haven: Yale University Press.

March, J. G. and Olsen, J. P. (1976). *Ambiguity and Choice in Organizations*. Bergen, Norway: Universitetsforlaget.

Martinet, B. and Marti, Y. M. (1995). *L'intelligence Economique: Les yeux et les oreilles de l'Entreprise*, Les Editions d'Organisation.

Merchet, J. D. (1997). La guerre economique, un art qui s'enseigne, *Liberation*. Paris.

Nash, L. L. (1981). Ethics without the sermon. *Harvard Business Review*, November-December.

Porter, M. E. (1980). *Competitive Strategy: Techniques for Analyzing Industries and Competitors*. New York: Free Press.

Simon, H. A. (1947). *Administrative Behavior: A Study of Decision-Making Processes in Administrative Organization* (First ed).

Simon, H. A. (1955). A behavioral model of rational choice. *Quarterly Journal of Economics*.

Simon, H. A. (1957a). *Administrative Behavior: A Study of Decision-Making Processes in Administrative Organization* (Second ed.).

Simon, H. A. (1957b). *Models of Man*. New York.

Simon, H. A. (1976). *Administrative Behavior: A Study of Decision-Making Processes in Administrative Organization.* (Third ed.). New York: The Free Press.

Thiètart, R. A. and Vivas, R. (1981). Strategic intelligence activity: The management of the sales force as a source of strategic information. *Strategic Management Journal*, 2(1), 15-25.

Turner, S. (1991). Intelligence for a new world order. *Foreign Affairs*, 70(4), 151-152.

Weick, K. E. (1976). Educational organizations as loosely coupled systems. *Administrative Science Quarterly*, 21, 1-19.

Weick, K. E. (1995). *Sense-making in Organizations*. Newbury Park, CA: Sage.

Chapter 17

Strategic and Ethical Issues in Outsourcing Information Technologies

Randall C. Reid
University of Alabama, USA

Mario Pascalev
Bank of America, USA

INTRODUCTION

Outsourcing of information technology (IT) is the transfer of a company's information technology functions to external vendors. Ordinarily, such transfer is considered only with regard to its strategic and economic impact on the organization. However, as the recent practice demonstrated, cost-benefit considerations and other strategic considerations are not sufficient to analyze an outsourcing case. Important ethical concerns relating to fiduciary responsibilities, insiders' bidding for outsourcing contracts, and the like, are also pertinent to the analysis of outsourcing.

This chapter will identify major ethical problems and will propose guidelines for ethical conduct in the process of outsourcing IT. Such guidelines could have broad practical implications for the practice of outsourcing.

The chapter will analyze literature on outsourcing models and professional ethical standards. It will have the following structure. First, the benefits and models of outsourcing information technology will be discussed. Second, ethical literature in general and professional organizations' codes of ethics in particular will be considered. Third, a recent case of IT outsourcing will be presented and analyzed. The ethical standards established in the thesis will be applied to the case. Finally, generalized ethical guidelines will be suggested for outsourcing models.

Previously Published in *Social Responsibility in the Information Age*, edited by Gurpreet Dhillon,

LITERATURE REVIEW: OUTSOURCING MODELS

Outsourcing of information technology (IT) is the transfer of a company's information technology functions to an outside agency. The first instance of IT outsourcing was the outsourcing of payroll processing. Many big companies have or are currently engaged in outsourcing (see Table 1). For some, IT outsourcing proved to be a viable strategy; for others, it was only a big headache.

Outsourcing has become a management strategy. According to a Forbes editorial (Sept. 22, 1997), "By 2000, 75% of enterprises will employ selective IT outsourcing as a routine means to increase competitiveness or gain new resources and skills." Nam et al. (1998, pp. 104-129) prompted a number of research studies and analyses of the practice. The scholars sought a conceptual answer to the question when to outsource IT and when not. In the next section, models of IT and their anticipated outsourcing benefits will be systematically reviewed.

General Management

In the area of general management, one of the important benefits of IT outsourcing is that it frees the company to concentrate on its core business competencies, rather than dealing with something which is unfamiliar and complicated. An additional advantage is in the simplification of the general managers' agenda.

There are also disadvantages of outsourcing IT to general management. As a result of the outsourcing arrangements, management loses flexibility. Most contracts are for 5-10 years. For the duration of the contract, the company is not in a position to change its IS strategy.

Finance

In the area of finance, the most important benefit of IT outsourcing invariably cited in the literature is that the company realizes various savings. The savings are

Table 1: Prominent companies that have outsourced IT

Company	Contract Value	Duration	Vendor
Del Monte Foods	$150 million	10 years	Electronic Data Systems
Health Dimensions, Inc.	$20 million	5 years	Integrated Systems Solutions Corp.
J.P. Morgan	$20 million	5 years	BT North America, Inc.
Signetics, Inc.	$100 million	10 years	Electronic Data Systems
U.S. Dept. of Housing	$526 million	12 years	Martin Marietta Corporation and Urban Development
General Dynamics Corp.	$3 billion	10 years	Computer Sciences Corporation

Source: Cheryl Currid and Company, Computing Strategies for Reengineering Your Organization (Rocklin, CA: Prima Publishing, 1994) 135.

due to the economies of scale ensuing from the specialization and larger volume of services provided by the outsourcer. Another financial benefit is the liquification of assets. Typically, it is a part of an outsourcing contract that the vendor purchases the computer hardware of the customer. Finally, the outsourcing transforms fixed costs into variable costs. Under an outsourcing arrangement, a company could reduce its costs by reducing the level of IT activity.

It is also maintained that the outsourcing vendor holds tighter control over supplies, such as paper, toner, and the like. The whole attitude toward the consumption of IT services becomes more frugal, as the soft dollars paid for the services of the internal IT department turn to hard dollars paid to an outside party. Theoretically, it is possible that all of the cost savings realized via outsourcing IT are less than the benefits of putting one's own IT "house" in order.

Personnel

There are certain benefits from outsourcing in terms of personnel. In the present job market for IT staff, it becomes increasingly difficult to attract or to retain good talent. Outsourcing arrangements supposedly give the company access to top talent in the field at affordable costs.

Additionally, under selective outsourcing arrangements, the company could keep its own IT personnel and use it only for specific projects and IS functions where their talents are best utilized. The routine support operations could then be transferred to the outside vendor.

A definite disadvantage of IT outsourcing is the high level of anxiety and low morale resulting from the news of imminent outsourcing. The best people may leave rather than being subject to uncertainty or undesired transfer to the vendor's payroll.

Technology

In the area of technology, the most important benefit of IT outsourcing is the instant access to new technology. Outsourcing provides a quick transition to the latest technology.

There are also some dangers of outsourcing IT associated with the technological function. Importantly, the economies of scale may make the vendor unable to move quickly to new technology. The emergence of industry-specific outsourcers "could result in the spread of systems' mediocrity throughout the industry because of the lack of new ideas and product differentiation" (Grover et al., 1994, p. 100).

There are three distinctions between kinds of benefits that evolved from the above list (and will be further elaborated on below). First, some benefits in the accounting sense may not be benefits in the economic sense. The economic benefit of IT outsourcing is only the difference of the cost to provide IT services in-house while operating efficiently and the cost to purchase the IT services from a vendor while the vendor is operating efficiently. If the latter cost is higher or equal to the former, outsourcing does not make economic sense. The cost difference between providing IT in-house while working inefficiently and outsourcing these services may result in accounting savings, but not in economic savings. In fact, the same or better results could come from simply organizing the in-house IT department more efficiently. The inefficient operation of IT services imposes opportunity costs on the company.

Another distinction that can be drawn between economic benefits is ones that are legitimate and ones that are ethically or morally questionable. It may be more cost-effective to win a contract by bribing an official rather than offering the lowest bid. Nevertheless, such conduct is unacceptable. A third useful distinction suggested by the earlier discussion is that between strategic and nonstrategic benefits of outsourcing IT. This distinction responds to the objection raised before that some benefits of outsourcing IT are not commensurate.

Strategic Discrepancy Model

Theng et al. (1994, 75-103), in an article titled "Strategy-Theoretic Discrepancy Model," argue that the decision to outsource information technology is based purely on strategic considerations (as opposed to accounting for various benefits). In practice, only the technological factors listed earlier are taken into account. They are evaluated in terms of their adequacy to the purposes of the organization. Teng et al. put forth three concepts. First, information technology is considered more generally as the capacity to manage information for strategic advantage. Further, strategy is defined as the match of internal organizational resources and the environmental opportunities and risks in view of the goals of the organization. Lastly, the authors propose that outsourcing will take place when there is a perceived gap between internal IT resources and the requirements of the environment. These concepts will be presented and discussed in turn.

First, Teng et al. propose a generalized view of information technology, shifting emphasis from technology and systems to information. They propose that IT management shifts from "a focus on technology to a focus on better information utilization and management, that leads to performance improvements and competitive breakthrough." Thus, IT is no longer simply one asset among the assets. It is information, a key asset, and indispensable for a firm's performance.

Other authors also support the strategic discrepancy model. After giving the received list of benefits and costs of IT outsourcing, McFarlan and Nolan also argue that the core of the outsourcing decision is in the strategic relevance of IT to a company at a particular time (McFarlan & Nolan, 1995, p. 15). Even though it is not as analytically rigorous as Teng et al., McFarlan and Nolan's paper has the advantage of offering a good practical decision-making tool. Current strategic

Figure 1: Strategic grid for information resource management

Current Dependence on Information	LOW ← Importance of IT for Competitive Advantage	→ HIGH
HIGH	Factory-uninterrupted service-oriented information resource management *Outsourcing Presumption:* Yes, unless company is huge and well-managed Reasons to consider outsourcing: • Possibilities of economies of scale for small and midsize firms. • Higher quality service and backup. • Management focus facilitated. • Fiber-optic and extended channel technologies facilitate international IT solutions.	Strategic information resource management *Outsourcing presumption:* No Reasons to consider outsourcing: • Rescue an out-of-control internal IT unit. • Tap source of cash. • Facilitate flexibility. • Facilitate management of divestiture.
LOW	Support-oriented information resource management *Outsourcing Presumption:* Yes Reasons to consider outsourcing: • Access to higher IT professionalism. • Possibility of laying off is of low priority and problematic. • Access to current IT. • Risk of inappropriate IT architectures reduced.	Turnaround information resource management *Outsourcing presumption:* No Reasons to consider outsourcing: • Internal IT unit not capable in required technologies. • Internal IT unit not capable in required project management skills.

Source: F. Warren McFarlan and Richard L. Nolan, "How to Manage an IT Outsourcing Alliance" (Palvia and Parzinger, 1995, p. 16).

relevance of IT to a company is plotted against the future importance of information resource management. The result is a "strategic grid" that could be used as a decision-making tool. (See Figure 1.)

As can be observed in Figure 1, four different strategic orientations are possible depending on present and future reliance on IT with their respective presumptions concerning outsourcing.

LITERATURE REVIEW: ETHICAL LITERATURE

Classical Approach to Ethics

Since ancient times, people have striven to come up with a way to tell right from wrong in a nonarbitrary way. The classical approach to this determination is to find principles founded in nature or ones that are self-evident and deduce rules for evaluation of conduct using correct reasoning (logic). As a result of this approach, a huge body of literature has arisen. Most influential streams in normative ethics today are the consequentialist (utilitarianism) and deontological (Kantian) ethics. Also prominent today are virtue ethics, feminist ethics of care, ethical egoism and contractarianism.

Utilitarianism maintains that we should judge the ethical merit of an action by the consequences it produces. Particularly, an action is good to the extent to which it tends to promote the greater happiness for the greater number of people. An action is bad to the extent to which it produces the opposite of happiness. The most important utilitarians are Jeremy Bentham and John Stuart Mill (Mill, 1979). Utilitarianism has a great intuitive appeal. It is predominant in contemporary health-care debate, allocation of scarce medical resources, organ transplants and the like. The notion of quality-adjusted life years is an elaboration on utilitarian ideas.

Kantian ethics maintains that an action is good to the extent to which it is done out of pure goodwill (i.e., done entirely out of good intentions). The most famous representative of this vein is the German Immanuel Kant. He believed that all rules of conduct should be tested against the so-called categorical imperative. Everybody should desire that a rule such as "Thou shalt not kill!" becomes a universal rule, without creating a logical contradiction.

It is unfortunate that ethical theories have adopted principles that contradict each other at a very fundamental level. It seems that we are left with little to do besides either disregarding the conclusions of ethical theories (anything goes), or stop worrying about fundamental principles and just focus on practical rules of ethical evaluation.

Rawls' Decision Procedure for Ethics

American philosopher John Rawls suggests a way of thinking about ethical evaluation which entirely sidesteps the issue about the first principles. Rawls asks: "Does there exist a reasonable decision procedure which is sufficiently strong, at least in some cases, to determine the manner in which competing interests should be adjudicated, and in instances of conflict, one interest given preference over another...." (Rawls, 1957, p. 177). The focus is shifted from foundational issues such as whether objective moral values exist, whether moral judgments are based on emotions, or whether autonomy or pursuit of happiness is the defining characteristic of human nature. The focus now is on the issue of whether a reasonable procedure for validating moral rules is available.

Rawls contends that such a decision procedure for ethics is available. "Competent moral judges" should be able to adjudicate in cases of conflicting interest and come up with valid ethical rules. The competency of these judges is defined as follows: The judges could be any people who know about the world what an average intelligent person would know. They are reasonable (i.e., they reason logically and are aware of their intellectual and moral predilections and take them into account when they consider particular cases). The moral judges are compassionate to the moral interests which are represented in the particular cases (Rawls, 1957, pp. 178-182).

The judgment rendered by the competent judges also is subject to a number of requirements. The judges should be immune from all foreseeable consequences of the judgment. The judgment adjudicates over actual conflicts of interest, as opposed to hypothetical cases. Further, the judgments are stable (i.e., at other times and places the competent judges would have arrived at the same conclusions about the same cases). Finally, the judgments should be intuitive rather than determined with a conscious reference to ethical principles.

The ingenious decision procedure for ethics proposed by Rawls is a very useful normative decision-making tool. The battleground of controversy about first principles, the nature of moral judgments, and the like is skillfully removed from the more pragmatic discussion of what is the right or wrong thing to do in a particular case. Rawls' procedure supplies professional ethics with a conceptual tool for adjudication in cases of conflicts of interest and for justification of moral rules.

Professional ethics goes through the throws of emancipation from academic philosophical ethics with its overwhelming conceptual heritage. The form it usually takes is paying lip service to one of the classical theories and moving on to reference of generally accepted social and moral values. The reason that professional ethicists would like to refer to classical theories is the need for validation of the generally accepted rules. The importance of the Rawlsian decision procedure is that it provides justification for the ethical rules based simply on the appropriateness of the procedure and not on the foundational justification of principles.

One important implication of the adoption of the Rawlsian decision procedure is the support it grants to the codes of professional ethics. Given that the rules in professional codes of ethics are subject to the decision procedure (i.e., their authors are intelligent, reasonable, and compassionate, and the judgments are impartial, stable, and intuitive), then these codes could be considered collections of valid ethical rules.

Testing the codes of professional ethics for the appropriateness of the decision procedure used for their creation is a major undertaking. Such a testing is clearly beyond the scope of the present inquiry. In order to simplify the task, it will be assumed that if distinct peer groups accept a rule independently, then it approximates a valid rule according to the decision procedure. This will leave us with somewhat tentative results, but one will be able to outline some important implications for the outsourcing model.

Outsourcing IT and Ethics

Potential Conflicts of Interest in the Context of Outsourcing IT

The types of potential conflicts of interest in the context of outsourcing include the interest of the following agencies: corporation customer (artificial person), corporation vendor (artificial person), agents of the customer, agents of the vendor, employees of customer, and general public. Potential conflicts of interest could arise between agents of the customer and their principal--the corporation customer (duties of loyalty and fiduciary duties), the agents of the vendor and their principal (duties of loyalty), customer and vendor (obligation to fulfill promises, loyalty to customer), and corporation customer and its employees (promises).

Professional Organizations' Codes of Ethics and Standards of Conduct

Although there is more than one way to present an ethical argument, the way proposed by John Rawls in *A Theory of Justice* has particular merit for its intuitive appeal. Ethical justification, according to Rawls, is an argument addressed to those who disagree with us or to ourselves when we are of two minds. To justify a moral conception to someone is to give him a proof of its principles from premises that we both accept. There are two elements in this method. The first element is the existence of mutually recognized starting points, that is, some consensus. The second element is the use of logical proof, or establishing logical relations between propositions. The appeal of this way of ethical justification is that it does not require subscribing to grand theories. It merely requires some agreement among reasonable persons.

Earlier, in the discussion of the decision procedure for ethics, it was determined that we could treat the codes of professional ethics as an approximation of valid ethical rules as long as these rules are arrived at independently by distinct professional communities. This framework will be used to research the professional codes of ethics for rules guiding the cases of conflicts of interest pertaining to outsourcing IT. If and when such rules are identified, they will be applied using deductive reasoning to analyze a particular outsourcing case.

There is a general consensus concerning agent-principal relations, loyalty and fiduciary duties in the DPMA's Code of Ethics, the Draft-Software Engineering Code of Ethics, the ACM's Code of Ethics, the IPG Society Code of Ethics, and the European Informatics Skills Structure Code of Professional Conduct. Particularly, these codes of professional ethics agree that an agent should refrain from "misrepresenting and withholding information" from his or her employer (http://courses.cs.vt.edu~cs3604/lib/WorldCodes/DPMA.Standards.html). He or she must "avoid or disclose any conflict of interest, which might influence his/her actions or judgment." He or she must "have no financial interest, direct or indirect, in any materials, equipment, hardware, or commercial software used by his/her employer

... unless he informs his/her employer in advance of the nature of the interest." An agent should refrain from using "resources of... employer for personal gain." He or she should "avoid conflicts of interest." The Draft Software Engineering Code of Ethics also speaks on the principal-agent conflict of interest. The IT professional must "promote no interest adverse to their employer's without the employer" (http://computer.org/standards/sesc/Ethics/Code.html) [4, 4.09].

With regards to the general societal interests, the agent should not take advantage of the lack of information... on the part of others for personal gain. The IT specialist shall not give opinions or make statements on professional programming projects that are inspired or paid by private interests. In the same vain, the professional shall act to correct or report any situation which could cause losses, whether humanly injurious or financially damaging. He or she should not promote their own interest at the expense of the profession, client or employer (Ibid. 6.04). The latter rule references the harm to society by such activities as insiders' trading.

The general consensus which we discover in the various reviewed codes of ethics suggests that members of the IT profession independently arrive at a core set of shared rules of conduct. Although only an approximation of the decision procedure for ethics, these rules pass the Rawlsian test. Therefore, it is established that there exists a set of rules of ethics against which to evaluate moral conduct in the IT field. Particularly, it is established that the codes of professional ethics of the relevant peer organizations such as DPMA and ACM contain rules condemning breaches of loyalty and fiduciary duties, lying, theft and conspiracy to cheat society out of fair competition. In the following section, a recent case of outsourcing IT will be presented and its implication for the ethics of outsourcing IT will be discussed.

A CASE OF OUTSOURCING INFORMATION TECHNOLOGIES

In this section, a recent fraud case of outsourcing will be discussed. Special consideration will be given to the ethical issues emerging in the case. Three top executives of a company producing a component for the automobile industry, namely the CEO, the executive vicepresident of technology and the executive vicepresident of operations, induced the director of management information systems to create a company to which the IT functions were outsourced. The profits were allegedly funneled back to consulting companies headed by each of these individuals.

The ethical merits of the case are considered from the perspective of widely shared ethical beliefs. The conditions for ethical justification are applied to the case. First, this is the presence of agreement on at least some basic principles and, second, this is the acceptance of the logic rules of derivation.

The actions of the executives violate beliefs about right and wrong which are widely accepted by the computer and information technology profession. Ultimately, these actions violate basic moral standards of the society. At least four elements of the executives' behavior appear to be unethical. The ethical violations include withholding material information from the employer, disregarding and harming the interest of the employer for one's own gain, misrepresenting the facts regarding future savings, unfair competitive practices based on insider information, and imposing undue hardships on the employees of the Company without appropriate justification.

The Company is an automobile manufacturing supplies firm, a subsidiary of an international parent company. In the beginning of 1990s, the Company experienced problems with excessive administrative overhead costs, including increased information technology costs. The parent company chairman repeatedly called for cost cutting. All functional areas were scrutinized for savings possibilities. The IT department, which had a reputation for lagging behind schedule and cost overruns, was particularly vulnerable.

In the early '90s, the IT staff consists of 40 application developers and 40 people associated with support and operations. The information systems were being run on an IBM mainframe computer. There were some microcomputers located at the plants that were used primarily for operations control. The software used was either written internally or bought off-shelf and customized. The expertise of the IT staff was primarily in COBOL and similar languages.

As the computer industry was moving away from mainframe computing towards server/client and UNIX environments, the Company's systems would require major revamping. Both systems and IT staff skills were becoming obsolete. The top management considered outsourcing of the IT function in order to solve the looming problems. It is a legitimate possibility that outsourcing was the best way for the Company to go. The problem is therefore not with the outsourcing decision per se. Rather it is with the motives of the decision makers and the ways they used to achieve it.

Three executives of the Company, its president and two executive vice-presidents, induced the director of management information systems to create a firm that will be referred to as IT Vendor, to which the Company outsourced its information technology (IT) functions for a period of 10 years. The outsourced functions include all mainframe processing services, personal computers and technical support. The new firm would supposedly reduce the projected IT costs of the Company from 72.4 to 52.8 million.

The particulars of the outsourcing agreement include that the Vendor was to maintain the same level of service, however, all orders would be fulfilled in time. The IT department was to be reduced to five managers, serving as liaisons with the

Company. The Vendor would hire some of the IT personnel. A third party would operate the data center.

The CEO and the two vice presidents accepted the proposal without seeking alternative bids for the job. The agreement was signed without publicity and was approved by the board of directors. Seventy-five people were immediately fired. Things improved in some ways, for instance, some projects were completed faster. Backlog of projects still existed because of the cuts. Also, the managing of the communication between the Company and the Vendor was hard for the five remaining managers. Three of them left their jobs by the end of the year. The costs steadily overran the projections. Finally, the outsourcing contract was terminated as part of legal action against the top executives. Besides the outsourcing deal, there were two other alleged fraud schemes including the sale of natural gas wells and leasing agreements.

Application of Ethical Standards to the Case

There is nothing morally controversial in outsourcing of IT per se. Outsourcing is based on prudential considerations of cost-effectiveness, access to expertise and technology, and competitiveness. The decision about outsourcing of IT depends on the particular circumstances of a firm, the type of industry, the importance of control and security of data, and the like. The ethical issues do not relate to the practice itself. Rather they relate to the way the practice is conveyed. Five types of ethical violations have been identified in the actions of the former executives. These are harming the interest of the employer in order to further one's own interest, withholding information crucial for decision making, misrepresenting the facts (lying) regarding future savings, unfair competitive practices based on insider information, and imposing undue hardships on the employees of the Company without sound economic program. The actions of the former executives violated the standards of ethical behavior generally accepted in the society and in the ethical theory.

Putting personal interest higher than the interest of the employer: The Company, supposedly because of the influence of the group, did not solicit alternate bids. The presence of more than one bidder would have revealed the costs of providing the service better.

The proposal did not offer access to new computer or managerial talent, as the new firm hired some of current management and most of the technical personnel. The new firm didn't offer access to new technology or expertise. Technically, the only two relevant outcomes were the transfer of ownership and control and the reorganization of existing resources. The former benefits only the executives and constitutes outflows for the Company. The latter might benefit the Company.

If we assume that the proposal sincerely advertised cost reductions through reorganization, these could not come from better-trained employees or the use of the advanced technology and expertise of an existing outsourcing firm. None of these could have been provided under the agreement. Savings could be expected only from areas such as business reorganization (improved managerial practices or layoffs). These means, however, were available to the management even before the outsourcing agreement. Avoiding a conflict of interest would require that the management pursued the reorganization within the Company, without claiming transfers to them.

Disloyalty: If the managers were sincere about reduction of costs, then they knew the reasons for high costs already. It was disloyal to withhold knowledge and information material for the performance of the firm. If, to the contrary, the managers did not believe that these means will really work, then they engaged in outright fraud. Of course, the managers might have both defrauded the Company and withheld potentially important information.

Misrepresentations of facts: Further developing the reasoning in above, we have to understand that the managers misrepresented facts when they suggested the amount of savings from the outsourcing deal. The deal looked attractive only because the current situation was poorly handled. It is not clear how much things would improve if the Company made a serious effort to reduce costs and improve the organization of its IT division from within. In any case, it would have been better that the current situation of mismanagement. The managers hid the issue of the baseline of comparison and presented the potential of self-improvement of the organization as a benefit which the new company contributes.

Obstructing fair competition: By using inside information to make an attractive offer to the Company, the managers deprived the companies providing IT outsourcing services of a fair chance to offer their bids and to reduce the Company's (and societal) costs. A bid like this is similar in nature to inside trading. It prevents fair competition, creates market imperfections, and increases societal costs.

Harming employees without redeeming reasons: The process of outsourcing involved layoffs, relationships of fear, and resentment among the affected employees. Causing emotional distress is only acceptable if it is dictated by offsetting reason. If I interpret the case correctly, and the major consideration was the personal gratification of the executives, then the suffering of the employees was unjustified.

Conclusions from the Case

Based on widely shared ethical beliefs, the actions of three top executives in outsourcing the IT department were morally wrong. There are at least four elements of their behavior to be unethical. The ethical violations include withholding material information from the employer, creating a conflict of interest, misrepresenting the future payoffs of the outsourcing agreement, unfair competitive practices based on insider information, and imposing undue hardships on displaced employees of the Company without justification.

The case study demonstrated that cost-benefit considerations are not sufficient to analyze an outsourcing case. Important ethical concerns relating to fiduciary responsibilities, insider bidding for outsourcing contracts, and the like are also pertinent to the analysis of outsourcing.

Seeking higher profits is a legitimate business goal. However, ends do not justify the means. Higher profits do not warrant ethically or morally questionable means. Under the pretense of seeking the benefits of IT outsourcing for their company corrupt agents may engage in fraud or breach of duties of loyalty. Establishing procedures of openness, solicitation of multiple bids, and the like will prevent companies from slipping into ethical controversies, costly lawsuits, and loss of face and reputation.

IMPLICATIONS OF THE STUDY FOR OUTSOURCING MODELS

The analysis of the outsourcing case demonstrated that ethics is an important consideration in an outsourcing decision. The practice of outsourcing sets the stage for a number of potential ethical conflicts, importantly, breach of trust and loyalty and breach of fiduciary duties. These could take place between the principal and its agents at the outsourcing customer, between the principal and its agents at the vendor, between customer and vendor, and between employer and employees. Agents could enter self-dealing schemes to the detriment of their company, or a company could use outsourcing to get rid of long-term commitments (i.e., retirement benefits) to employees. There may be other types of ethical conflicts, but they are not prominent in the present case.

The most important conclusion from the present analysis is that when a company considers outsourcing, it should take into consideration potential ethical conflicts. Perhaps, it is enough to say that acts that are ethically wrong should not be done. Still, allowing unethical behavior on the part of the company or its agents does some very real harm. These include, on the company level, dwindling of the morale of the employees, resignations of important employees who would not like to be a part of the disgraced company, loss of respect for the management of the company, and distrust of the mission and objectives of the company. On a societal scale, there could be costly lawsuits and loss of the goodwill of the public, as well

Figure 2: Ethical issues on the strategic grid

Current Dependence on Information (vertical axis: LOW to HIGH); Importance of IT for Competitive Advantage (horizontal axis: LOW to HIGH)

	LOW importance of IT for competitive advantage	HIGH importance of IT for competitive advantage
HIGH current dependence on information	Factory-uninterrupted service-oriented information resource management *Outsourcing Presumption:* Yes, unless company is huge and well-managed Considerations in outsourcing: • Possibilities of economies of scale for small and midsize firms. • Higher quality service and backup. • Management focus facilitated. • Fiber-optic and extended channel technologies facilitate international IT solutions. ▪ Vendor's vagaries.	Strategic information resource management *Outsourcing presumption:* No Considerations in outsourcing: • Rescue an out-of-control internal IT unit. • Tap source of cash. • Facilitate flexibility. • Facilitate management of divestiture. • Embezzlement. • Vendor's disloyalty. • Company's disloyalty to employees.
LOW current dependence on information	Support-oriented information resource management *Outsourcing Presumption:* Yes Considerations in outsourcing: • Access to higher IT professionalism. • Possibility of laying off is of low priority and problematic. • Access to current IT. • Risk of inappropriate IT architectures reduced. • Vendor's vagaries.	Turnaround information resource management *Outsourcing presumption:* No Considerations in outsourcing: • Internal IT unit not capable in required technologies. • Internal IT unit not capable in required project management skills. • Embezzlement. • Vendor's disloyalty.

as loss of trust by business partners. Most importantly, there will be misallocation of economic resources.

It is suggested that the models for evaluation of outsourcing decisions should include the ethics factor into their consideration. The reader recalls that the strategic discrepancy model is preferable to the list of benefits model in that it identifies the key strategic impact of an outsourcing decision for a company, rather than considering factors of varied degrees of relevance. The strategic interaction model was preferred to the strategic discrepancy model in its inclusion in the analysis of the interaction with the vendor. It is important to understand that the goal of this chapter is not to determine the best model for evaluating outsourcing decisions. It is not necessary to accept the strategic interaction model in order to accept the points about the need for ethical concern in outsourcing. The conclusions of the

Figure 3: Ethical dimension in the interaction between a customer and a vendor

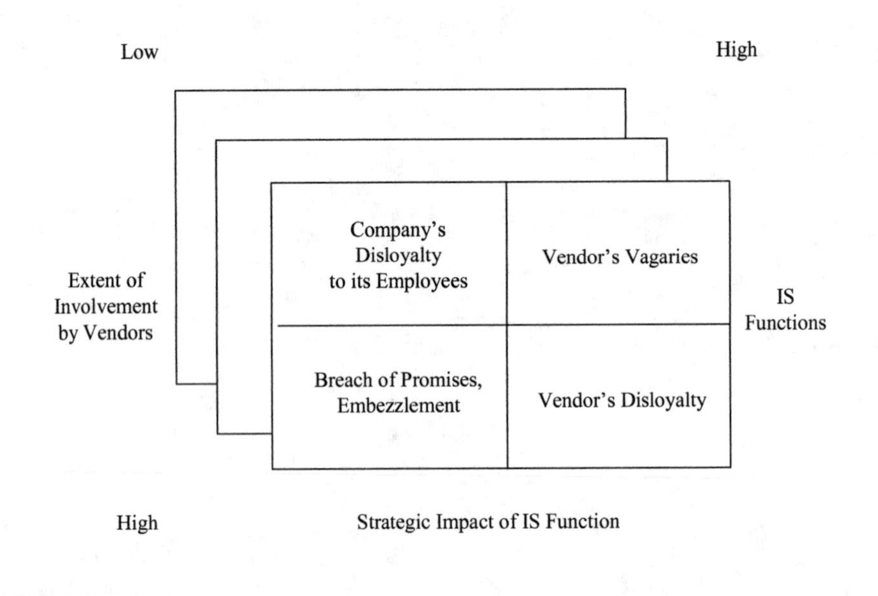

analysis are available for the adherents of all three models. Briefly, each of the models will be improved if it includes ethical analysis of the particular decision.

List of Benefits and Ethics

According to the logic of this model, the absence of ethical concerns will be listed as a benefit of a particular outsourcing deal. The potential and actual ethical concerns will be listed as disadvantages of the outsourcing deal. Loss of the goodwill of the community will be considered as a disadvantage in the area of general management. The dwindling of morale, resignations, and detriment of leadership are disadvantages in the area of human resources management. The model warns against these but, as stated before, it leaves it open whether an instance of unethical behavior has more weight in the mind of a manager than some petty savings.

Strategic Discrepancy and Ethics

This model would make even more adequate use of the ethical analysis. It recognizes that some factors are greatly more important than others for the company, as they affect its strategic goal. The decision to adhere to the professional code of ethics is a strategic decision. Internally, it affects the human resources of the company. If the company is loyal to its employees and the agents are loyal to their principal, that keeps the morale high and protects the authority of the management.

Externally, the decision affects the overall public perception of the company and creates goodwill. Insider trading and obstruction of fair competition are activities that would generally antagonize the public and will produce a negative reaction (legal action, regulation, or simply a negative image). Including ethical issues to be considered in each strategic quadrant could enhance McFarlan and Nolan's strategic grid presentation (see Figure 2).

Strategic Interaction and Ethics

The final model also would benefit from the inclusion of ethical considerations. Everything said about the previous model applies here. The decision to adhere to ethical standards is a strategic one. It influences the internal management of resources, as well as environmental factors. The model is represented graphically as a two-dimensional grid which maps the extent of involvement by vendors against the strategic impact of an IS function. Then the functions are projected on vertically, forming a third dimension. The ethical analysis could appropriately be depicted on one of the functional cards (see Figure 3).

The diagram suggests that we could classify ethical conflicts in the context of outsourcing according to the level of involvement of the vendor and the strategic role of IT technology. When the vendor's involvement is low, the potential for ethical damage is within the company-employees relationships. The anticipated ethical issues stemming from the vendor will be limited to vagaries, lack of commitment, and the like. As the involvement of the vendor increases, considerations such as embezzlement become more prominent. When the vendor's involvement is high and the IT function is strategic for the company, there exists a potential for vendor's disloyalty, such as betraying secrets of the company.

At a more conceptual level, first, the diagram brings home the idea that ethical conflicts are a strategic concern for a company involved in outsourcing. The higher the roles of the IT function, the bigger the impact of ethical conflicts. Second, the diagram emphasizes the interactive nature of outsourcing and the related source of ethical conflict. Ethical conflict could be internal for the company, or it could be embedded in the interaction between the company and the vendor.

CONCLUSION

Outsourcing of IT is the transfer of a company's information technology functions to external vendors. Such transfer is typically evaluated with regard to its strategic and economic impact on a company. However, as the analysis of a recent case demonstrated, ethical concerns relating to fiduciary responsibilities, insider bidding for outsourcing contracts, and similar ethical conflicts are also pertinent to the analysis of outsourcing.

The analysis identified major ethical problems and suggested a systematic way of evaluating ethical conduct in the process of outsourcing IT in the context of a company's strategy. The ethical considerations were incorporated in the most influential models of outsourcing: the list of benefits model, strategic discrepancy model, and strategic interaction model. The models were developed as to include a classification of ethical conflicts with regards to the level of the vendor's involvement and the strategic importance of the IT function for the company. Hopefully, this study will enhance the way companies look at the decision to outsource information technologies and provide practical guidelines for outsourcing IT.

REFERENCES

Forbes. (1997). Outsourcing to the rescue. *Forbes*, September, 1(1), 22, 25-26.

Grover, V., Cheong, M. J. and Teng, J. T. C. (1994). A descriptive study on outsourcing of information systems functions. *Information and Management*, 27, 32-44.

Kant, I. (1993). *Grounding for the Metaphysics of Morals*. Translated by James W. Ellington. 3rd ed. Indianapolis: Hackett.

McFarlan, F. W. and Nolan, R. L (1995). *Sloan Management Review*, 36(2), 9-23.

Mill, J. S. (1979). *Utilitarianism*. (George Sher, Ed.). Indianapolis, IN: Hackett.

Nam, K., Chaundhury, A., Rajagopalan, S. and Rao, H. R. (1998). Dimensions of outsourcing: A transaction cost framework. *Managing Information Technology*, 104-129.

Palvia, P. and Parzinger, M. (1995). Information systems outsourcing in financial institutions. In Khosrowpour, M. (Ed.), *Managing Information Technology Investments with Outsourcing*, 129-154. Hershey, PA: Idea Group Publishing.

Rawls, J. (1957). Outline of a decision procedure for ethics. *Philosophical Review*, 66, 177.

Theng, J. T. C., Cheong, M. J. and Grover, V. (1994). Starategy-theoretic discrepancy model. *Decision Sciences*, 25(1), 75-103.

Chapter 18

Ethics, Authenticity and Emancipation in Information Systems Development

Stephen K. Probert
Cranfield University, United Kingdom

INTRODUCTION

This chapter describes research in progress on the philosophical concept of authenticity – used as a framing device for providing an interpretation of aspects of both ethical and practical action on the part of information systems (IS) professionals. Ethical codes and prescriptive IS development methods for IS professionals can be found in most developed countries in the world. Here it is argued that ethical codes and IS methods may be of limited value in IS work. One key problem here is that IS analysts and designers have to *intervene* in organisations (and thereby intervene in the lives of the members of those organisations). It is argued that an important issue for IS research is whether they choose to do so in (what will be characterised as) an *authentic* manner, rather than doing so in sincere adherence with either a code of professional ethics or with a series of methodological precepts.

PERSONAL AUTHENTICITY AND ETHICAL CODES

Firstly, to characterise the concept of authenticity, a brief explanation will be given. Given that there is a lack of *absolute* guidance as to how one is to act in any given situation, the question of "what should one do...?" raises severe difficulties. Some sorts of authenticity questions may be familiar to the readers of this chapter. As a consultant, the author experienced several authenticity problems, a few are given as example questions here:

Previously Published in *Challenges of Information Technology Management in the 21st Century*, edited by Mehdi Khosrow-Pour, Copyright © 2000, Idea Group Publishing.

1. Should I use a methodology which has embedded values that I do not agree with?
2. Should I use a methodology which, in my judgement, is wholly inappropriate to the circumstances pertaining in the organisation?
3. Should I attempt to improve organisational performance by introducing greater accountability in a low-wage organisation?

These are difficult ethical questions, and whilst some of these may be covered by the codes of conduct and practice of professional IS bodies, others may not be (see Walsham, 1996). Also, such decisions require degrees of interpretation, and therefore judgements about such matters are likely to vary from person to person. In any case, not all IS professionals are members of professional societies, and not all those members may be aware of the codes of conduct and practice, and no doubt some will choose to ignore such things. More importantly, adherence to any such code is unlikely to be practically *enforceable*; adherence will therefore have to be "granted" voluntarily by the IS professionals concerned:

> "In the scientific community the medical specialist has better defined ethical codes than most other groups... They are also enforced by powerful sanctions such as expulsion from the medical profession if serious infringements occur. Many other professionals, including the British Computer Society, have also drawn up ethical codes but these are often vague and difficult to apply and enforce... Ethical responsibilities will also vary both with the nature of work that is being carried out and the nature of the social environment where the work is conducted." (Mumford, 1995, p. 6)

Because the value of ethical codes are limited, the sorts of questions characterised above (which all IS professionals must probably face from time to time) may best be understood as questions of *personal authenticity*, rather than being understood as strictly *ethical* questions. Indeed, it has been suggested that, "[T]he concept of authenticity is a protest against the blind, mechanical acceptance of an externally imposed code of values." (Golomb, 1995, p. 11)

The concept of authenticity is often primarily connected to considerations put forward by Nietzsche (1844-1900). Nietzsche's statements and concerns about such issues are a constant theme in his texts (especially 1956 and 1974). Cooper elaborates the concept of authenticity via some examples from teaching. He explicates the problems thus:

> "A familiar disturbance felt by the teacher arises when some of these [educational] policies, values, or whatever, are not ones to which he can subscribe... The disturbance produces a problem of authenticity, for unless the teacher resigns or is willing to invite considerable friction at work, he must simulate agreement to views that are not his. [Alternatively] ... The

thought which may strike the teacher is not that he cannot subscribe to, or authoritatively transmit, various beliefs and values, but that he has slipped into, fallen into, unreflective acceptance of them. They have become part of the school's furniture; they go with the job like the free stationery." (Cooper, 1983, p. 4)

Such questions are intensely personal, and researching how IS professionals deal (or should deal) with such questions as arise in IS practice will be necessary if real progress is to be made towards the aim of improving IS practice, because slavish adherence to externally imposed codes of conduct is not necessarily a guarantor of ethically proper behaviour (it has been argued).

AUTHENTICITY AND METHODOLOGICAL PRECEPTS

An example of a tension between methodological adherence and authentic systems development practice can be found within the ubiquitous concept of the *systems development life cycle*. This was originally derived from an empirical study by Barry Boehm (Boehm, 1976). The consequent life cycle model has been absorbed into nearly every structured IS method propounded ever since; if it is criticised, it is criticised as being a *prescription* that does not "work" in practice (whatever the precise form of the criticism takes). The usual criticism runs along the lines that the longer one takes to "get the requirements right" the longer it takes to develop a system at all - and the greater the likelihood becomes that the requirements are "out of date":

> "The criticisms that are periodically made of the development life cycle concept ... mostly focus on its being a linear, sequential model in which each stage must be completed before the next is begun. This means that it relies heavily on the initial definition of the problem being complete and correct and that the users' requirements will not change in the time taken to progress to final implementation. In the case of modern complex information-systems neither of these assumptions can safely be made ..." (Lewis, 1994, p. 75)

Nevertheless the widespread use of life cycle methods for IS developments continues relentlessly (although numerous alternative approaches are often propounded). A recent UK survey was conducted to investigate the use of systems development methods (amongst other things). This survey indicated, "Within systems development, 57% [of systems development staff] claim to be using a systems development methodology" (Fitzgerald et al., 1998). The effect of the widespread adoption of structured methods is to remove personal authenticity from the systems development personnel. Lewis argues:

> "The legacies of hard systems thinking, such as the idea of the development life cycle, have become so deeply ingrained in IS thinking that only rarely is note taken of the constraints that they impose upon the way we view the development of information-systems." (Lewis, 1994, p. 75)

Now, as received wisdom becomes a guiding force for decision making, so the possibilities for making any *genuine* decisions tend to evaporate. As Golomb argues:

> "In the context of our everyday humdrum lives, it is hard to know what we genuinely feel and what we really are, since most of our acts are expressions and consequences of conditioning, imitation and convenient conformity." (Golomb, 1995, pp. 24-25)

Adherence to methodological prescriptions may provide systems development staff with a convenient set of reasons for not doing what they (truly) feel that they *ought* to do. The point to stress here is that these motivations (to do what one ought to do on authentic versus methodological grounds) are not identical – they are very different. Indeed, Wastell has pointed out the degree to which the adherence to methodological prescriptions has a value as a defence mechanism for systems development staff (Wastell, 1996). Although the main focus of Wastell's paper is to demonstrate how it comes about that methodology gets used as a social defence mechanism he also argues that what is actually needed in systems development situations is quite different:

> "[M]any analysts apparently developed a fetsihistic dependence on methodology ... They appeared to withdraw from the real job of analysis, of engaging with users in an open and frank debate about system requirements. Instead they withdrew into the womb of security provided by the method[1] . They worried about details of notation, of whether the method was being correctly implemented and of the need to press on and fulfil deadlines rather than ensure that they had really understood what the users wanted." (Wastell, 1996, pp. 35-36)

This can be interpreted as a failure of authenticity on the part of the systems development staff encountered by Wastell.

AUTHENTIC INTERVENTION

Many models of authenticity have been propounded, but in this short paper we may consider the Nietzschean approach in isolation. Structured / life cycle methodological precepts make little allowance for the influence of choice on the part of the IS professionals - who will be (methodologically) guided to investigate practically everything relevant in a particular study. Of course, such detailed and thorough investigations not only difficult to achieve practically, but run counter to the actual social-psychological conditions in which analysts operate; primarily, on organisational (social) grounds:

> "The modern organisational environment is a far cry from the well-ordered world of the classical bureaucracy, with its elaborate hierarchical division of labour and highly routinized procedures. The modern organisation, in contrast, is characterised by constant innovation, by flux and fluidity [which] presents a potent challenge to the social defences that characterise the traditional organisation, such as the bureaucratic ritual, which contain anxiety by narrowing attention and by defining rigid roles. The new demands require a broadening of rules, wider boundaries, increased integration and interdependence." (Wastell. 1996, pp. 34-35)

Can the concept of Nietzschean authenticity help us to understand the psychological demands placed on contemporary IS professionals? Nietzsche's arguments on such issues can be found in Book Five of *The Gay Science* (Nietzsche, 1974). Golomb makes the following points – concerning how Nietzsche conceptualised the relationship between authenticity and epistemology - in a clear manner:

> "An individual's life comprises a boundless number of experiences and notions, including a tremendous amount of superfluous information. Through awareness of one's authentic needs one may organise and refine this chaos into a harmonious sublimated whole. Initially the self is a bundle of conflicting desires and an array of contradictory possibilities. The self's unity is a function of its own decisions and creations... The search for authenticity is seen as the wish to reflect one's own indeterminacy by spontaneous choice of one of the many possible ways of life." (Golomb, 1995, p. 69)

Prima facie, a great deal of systems development work in a turbulent organisational environment can – indeed must – depend on the authenticity of the development staff if good systems are to be developed. Slavish adherence to methodological prescriptions can only serve to deny the insights and wisdom attained by systems development staff (about the actual needs of the organisation) over many years of experience. Moreover, it can be conjectured that the widespread use of contract IS/IT staff – often with disastrous consequences (Currie, 1995) – is indicative that insufficient attention has been paid, by IS managers, to the role that authenticity plays in good systems development.

CONCLUSION

This chapter has reported work in progress on the philosophical concept of authenticity as providing both a better way of understanding the role played by ethical codes and IS methods, and as way of characterising actual IS practice, in modern organisations. Although Nietzsche's version of authenticity has been basically characterised, other philosophers and authors of literature have made important contributions to the debate – including Heidegger. Further research would need to be investigate these views also. Finally, it should be noted that Adorno (1973) provides a powerful critique of the whole notion of authenticity. Suffice it to say here that such a critique needs to be taken seriously and warrants further research.

ENDNOTE

[1] In case study reported in Wastell (1996), the method used was the UK's SSADM.

REFERENCES

Adorno, T. (1973). *The Jargon of Authenticity*, London: Routledge and Kegan Paul.

Boehm, B. (1976). Software engineering, *IEEE Computer.* C-25(12).

Cooper, D. E. (1983). *Authenticity and Learning*, London: Routledge and Kegan Paul.

Currie, W. (1995). *Management Strategy for I.T.: An International Perspective*, London: Pitman.

Fitzgerald, G., Phillipides, A. and Probert, S. (1998). Maintenance, enhancement and flexibility in information systems development, in J. Zupancic, G. Wojtkowski, W. Wojtkowski and S. Wrycza (Eds.), *Systems Development Methods for the Next Century, 2*, Plenum, New York.

Golomb, J. (1995). *In Search of Authenticity*, London: Routledge.

Lewis, P. (1994). *Information-Systems Development*, London: Pitman.

Mumford, E. (1995). Human development, ethical behaviour and systems design, in N. Jayaratna, R. Miles, Y. Merali and S. Probert (eds.), *Information Systems Development*, Swindon: BCS Publications.

Nietzsche, F. (1956). *The Birth of Tragedy and the Genealogy of Morals*, New York: Doubleday.

Nietzsche, F. (1974). *The Gay Science*, New York: Random House.

Walsham, G. (1996). Ethical theory, codes of ethics and IS practice. *Info Systems J.* 6: 69-81.

Wastell, D. G. (1996). The fetish of technique: methodology as a social defence. *Info Systems J.* 6: 25-40.

Chapter 19

On the Role of Human Morality in Information System Security: From the Problems of Descriptivism to Non-Descriptive Foundations

Mikko T. Siponen
University of Oulu, Finland

INTRODUCTION

The relevance of security solutions and procedures depends on the motivation of the users to comply with the security solutions/procedures provided. Many studies indicate that users fail to comply with information security policies and guidelines (e.g., Goodhue & Straub, 1989; Parker, 1998; Perry, 1985). It is widely argued (e.g., Loch & Carr, 1991; Anderson, 1993; Parker, 1998; Vardi & Wiener, 1996; Neumann, 1999) that a remarkable portion of security breaches are carried out by organizations' own employees. Several proposals have been made to tackle this human problem, the solutions range from 1) increasing the users' motivation (e.g., McLean, 1992; Perry, 1985; Siponen, 2000; Thomson & von Solms, 1998), 2) using ethics (e.g., Kowalski, 1990; Leiwo & Heikkuri, 1998a, 1998b), 3) organizational/professional codes of ethics (e.g., Harrington, 1996; Straub & Widom, 1984; Parker, 1998), to 4) using different deterrents (e.g., Straub, 1990). With respect to the second issue–Can human morality function as

Previously Published in the *Information Resources Management Journal, vol.14, no.4*, Copyright © 2001, Idea Group Publishing, and in *Social Responsibility in the Information Age*, edited by Gurpreet Dhillon, Copyright © 2002, Idea Group Publishing.

a means of ensuring information security?–the existing works can be divided into two categories. The first category covers expressions concerning the use of human morality including Kowalski (1990), Baskerville (1995), Siponen (2000) and Dhillon and Backhouse (2000):

- "Security administrators are realizing that ethics can function as the common language for all different groups within the computer community" (Kowalski, 1990).
- "Proper user conduct can effectively prevent [security] violations" (Baskerville, 1995, p. 246).

The second claims that the use of ethics is useless or, at best, extremely restricted (Leiwo & Heikkuri, 1998a, 1998b).

This chapter argues, following the scholars of the first category, that human morality has a role as a means for ensuring security. But to achieve this goal solid theoretical foundations, on which a concrete guidance can be based, are needed. The existing proposals (e.g., Kowalski, 1990; Baskerville, 1995; Dhillon & Backhouse, 2000) do not suggest any theoretical foundation nor concrete means for using ethics as a means of ensuring security. The aim of this paper is to propose a framework for the use of ethics in this respect. To achieve this aim, a critique of the relevance of ethics must be considered. The use of human morality as a means of ensuring security has been criticized by Leiwo and Heikkuri (1998a, 1998b) on the grounds of cultural relativism (and hacker ethics/hacking culture). If cultural relativism is valid as an ethical doctrine, the use of human morality as a means of protection is very questionable. It would only be possible in certain "security" cultures, i.e., cultures in which security norms have been established–if at all. However, the objection of Leiwo and Heikkuri (1998a, 1998b) is argued to be questionable. We feel that cultural relativism has detrimental effects on our well-being and security. Things might be better if the weaknesses of cultural relativism were recognized. This paper adopts the conceptual analysis in terms of Järvinen (1997, 2000) as the research approach. An early version of this paper was presented at an international conference on information security (IFIP TC11, Beijing, China, 2000).

The chapter is organized as follows. In the second section, the possible ethical theoretical frameworks are discussed. In the third section, the objections to the use of ethics as a means of protection based on cultural relativism (descriptivism) are explored. In the fourth section, an alternative approach based on non-descriptivism is suggested. The fifth section discusses the implications and limitations of this study. The sixth section summarises the key issues of the chapter including future research questions.

THEORETICAL FRAMEWORKS

Ethical Theories

The philosophical ethical theories can be classified into two categories, descriptivism and non-descriptivism (Hare, 1997). In this chapter, descriptive theories refer to ethical doctrines that attempt to draw a morally or action-guiding conclusion purely from a set of factual premises, such as prevailing cultural habits. In other words, the separation between descriptivism and non-descriptivism can be retraced to Hume's thesis that moral norms (what we ought to do) cannot be drawn from a set of factual matters. Those theories arguing that factual matters imply moral norms are called descriptivism, as opposed to non-descriptivism (see Figure 1). This simplistic division is chosen for a practical reason; it is perhaps the simplest classification and therefore helps us to understand the different theoretical possibilities available and their one fundamental difference.

We have left out religion-based ethical theories (e.g., Christian ethics) from the categorization. The reader is advised to look at Outga (1972) for more religion-based ethical theories and the question of desriptivism versus non-descriptivism ("is/ought"-problem). We believe Siponen (2001), Siponen and Vartiainen (2001), as many others have already proposed (e.g., Hare, 1981, 1997; Taylor, 1975), that descriptive theories such as cultural relativism and intuitionism are inadmissible as moral qualifiers. Instead of attempting to find what is morally right and wrong, descriptive theories, at best, pay lip service to prevailing cultural moral notions (cultural relativism) or individuals' intuitions (intuitionism). In the worst possible scenario, descriptive theories may be used as an excuse to indulge in morally questionable behaviour (e.g., Nazism or hacking), as shall be seen in Section 3. In

Figure 1: Depicts the division and some ethical theories

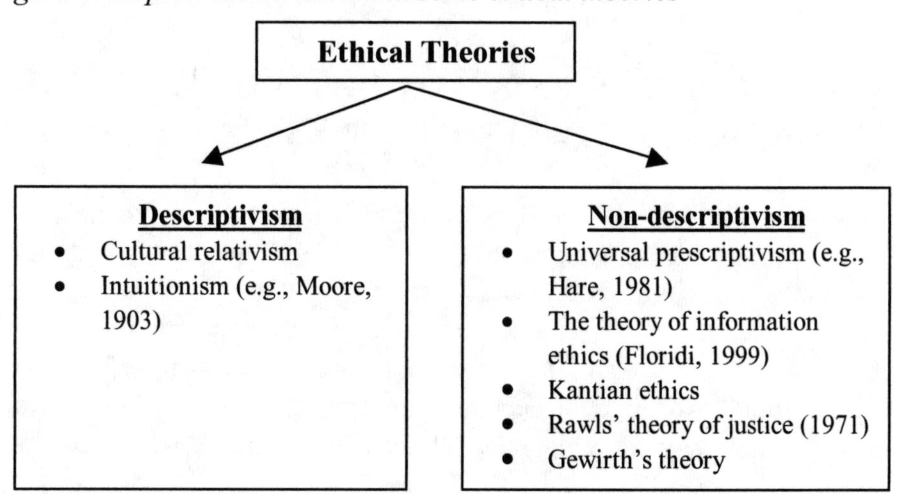

Section 4, it is proposed that we should look to non-descriptivism to provide solutions. In this study, the term moral means what people regard as right and wrong–how we should act in the final analysis. Ethics refers to moral philosophy, i.e., ethical theories discerning what is morally right and wrong.

Overriding Thesis

In order for human morality to be useful in security procedures, it is necessary that we should have an intrinsic sense of moral responsibility, in other words, a sense of duty forcing us to follow our moral concerns: to find out what is morally right. If all people were totally amoral (i.e., did not care what is morally right) or if theories such as cultural relativism were considered as valid moral qualifiers (as proposed by Leiwo & Heikkuri, 1998a, 1998b), human morality could not function as a means of ensuring security. We need to examine whether there is such a thing as "moral responsibility"? And, if there is, how strongly does it guide our behaviour? These two questions (and the relevance of human morality for security) can be retraced to the validity of the overriding thesis suggested by R.M. Hare (1963). Hare claims that moral concern overrides all other nonmoral concerns (overriding thesis). In other words, given that one regards unauthorized copying of software as morally wrong (moral concern), one should not copy software even if one would receive financial gain (nonmoral concern). Ladd (1982, 1989) has suggested a similar view: Our moral responsibility (what is morally right) overrides other forms of responsibilities. Smith (1984) even argues that it is more understandable to act for moral reasons than for nonmoral ones. This latter view has been criticised by Dancy (1994) on the grounds that moral, or justified, reasons do not imply motivation per se (Dancy argues that one may see nonmoral reasons as understandable as well). If Hare's overriding thesis is valid, appealing to human morality is crucial for ensuring security.

ANALYSIS OF THE EXISTING DESCRIPTIVE APPROACH AND ITS WEAKNESSES

Theoretical Underpinnings

Leiwo and Heikkuri (1998a, 1998b) argue that human morality cannot serve as a means of protection against security violations–particularly in a global environment (e.g., the Internet)–because of cultural relativism. They argue that moral values are subjective in the sense that they cannot be transferred from one place or moral system to another (Leiwo & Heikkuri, 1998b, p. 275). In other words, the morality of an action depends on culture: What is morally right in one culture may be morally wrong in another culture. This argument also involves ethical descriptivism, since they indicate that moral judgement has a truth value (e.g., true/false): "the truth values of ethical value systems . . ." (Leiwo & Heikkuri, 1998b, p. 275).

Leiwo & Heikkuri further engage in cognitivism, which is mainly an epistemological claim stating that values can be known to be true. In the case of cultural relativism, exploring the moral values of cultures is argued to validate this cognitivistic claim (values can be known to be true). Because it is regarded as a sociological fact that morality (what people do/consider as right and wrong) depends on culture, relativists claim that what every culture does is equally right or true. The reasoning of Leiwo and Heikkuri is similar to this. They argue that cultural relativism is valid and moral views differ with respect to information security. The culture of hackers and hacker ethics was provided as a proof (Leiwo and Heikkuri, 1998b, p. 275). Since "hacker ethics" by hacker "Knightmare" (see more Fiery, 1994) and cultural relativism were provided by Leiwo & Heikkuri as an example to indicate the inadequacy of human morality as a means of ensuring security, we next examine these concepts in more detail.

Weaknesses of the Descriptive Approach

Is-Ought Dualism

The approach of Leiwo and Heikkuri does not take into account the factual/normative dualism first recognised by David Hume. This dualism is also known as Hume's law or the thesis "no ought from an is." We share Popper's view that Hume's law is "perhaps the simplest and most important point in ethics" (Popper, 1948). "No ought from is" means that factual premises alone, i.e., "is" cannot imply norms, i.e., "ought" statements. So, for example, what people/culture regard as morally right ("is") does not provide the answer to what one morally ought to do ("ought"). Leiwo and Heikkuri (1998a, 1998b) fall into this ("is" implies "ought") fallacy by first observing "is" matters: hacker ethics/culture. From this they deduce that, due to relativism (what every culture does is right), the actions of hackers (or hacker ethics) are right per se, therefore they are implying an ought from an is.

Although there are attempts to prove the invalidity of Hume's thesis "no ought from an is," including Searle (1964), Gewirth (1974) and MacIntyre (1981), they do not serve as persuasive objections especially when addressing cultural relativism. For example, Searle's attempt to break Hume's law is widely criticised as being a game which can only be played if the players accept the rules of the game provided by Searle (Hare, 1964a). It may serve as an indication that it is possible to persuade someone to form an "ought" (or moral) judgement by giving "is" matters without "ought" matters. But does this prove that this kind of treatment would be desirable? The most difficult problem with Searle's argument is that "the rules of the game" are based on persuasion, and the rules do not provide any restrictions regarding the contents of the strategy agreed upon, which means that the game is, for instance, open to lying. The main objection to this is, if lying is acceptable, we are throwing ethics and morals out of the window. Gewirth's idea (of equality) is closely connected to the universality thesis, which serves as a basis for his ethical theory/sociopolitical theory (e.g., see Gewirth, 1978) and therefore, even if accepted, does not help cultural relativists. Moreover, Kohlberg's thesis is the opposite of cultural relativism as it accentuates Kantian universality thesis ("act only on maxims which you want to be universal laws") as the highest state of moral development.

Is-Ought Dualism and a Practical Example

The weakness behind the thesis of Leiwo and Heikkuri (1998a, 1998b) can also be considered by using a more down-to-earth example. An employee is working in a company involved in a top-secret project. The employee joins an association that has its own moral code and the company accepts their employee's joining this association because the company considers the activities of the association harmless. Later, the members of the association start to take an interest in philosophy and find out that their background has been forgotten and is rather different to the one they assumed was correct. They are advocates of cultural relativism and so they perform a "who are we really/what are our moral values" perusal (e.g., provided by Sandel, 1982). This is a way to reflect "is" matters, i.e., how things were/are, and to allow this to determine how things should ultimately be. As a result of this, they find a new moral code that better reflects their original background and they are positive that this is their real moral code. This new code allows hacking, the result being that the employee is now encouraged to break into his company's system (or otherwise allow the association access to top-secret information). If the company and the employee acknowledge cultural relativism as a valid moral qualifier, they have no (moral) right to either prevent such actions or take any stand with respect to these actions, otherwise they have interfered with the other culture.[1] Thus, as mentioned, any (moral) involvement with the other culture is not acceptable according to cultural relativism. The company in our example therefore cannot take any moral stand concerning some other culture (the association in our example). This is not a very convincing justification.

Hacker Ethics and Cultural Relativism

Knightmare states, "hacking is something that I am going to do regardless of how I feel about its morality" (Fiery, 1994, p. 162). It is difficult to see how such a view can have connections to the domain of moral discourse (or real hacker ethics). However, such evasions of moral reflection can be justified given that cultural relativism is regarded as a valid moral qualifier. Knightmare could insist that hackers form a culture and due to cultural relativism, we should allow them to do whatever they do. This illustrates another weakness of culture relativism (and hacker ethics, as well). Cultural relativism and Knightmare's version of hacker ethics do not truly attempt to discover what is morally right and wrong. Rather they avoid moral scrutiny and uphold dogmatism.

Hegel: Cultural Relativism and Hacker Ethics

Leiwo and Heikkuri (1998a, 1998b) also validate cultural relativism and hacker ethics on the basis of Hegel. This totally relativistic view is not shared by Hegel.[2] Hegel recognises the problems associated with cultural relativism. Cultural relativism holds that all beliefs/belief systems are equally true. Hegel does not share this view. Hegel sees that moral conflicts should be solved in such a way that one's freedom and above all the coherence of the state (i.e., government/country) are ensured (Sabine, 1963, p. 655). Hence, if one follows Hegel's doctrine when considering if it is acceptable to allow hacking, one needs to ponder which alternative ensures the coherence of the state. Note that hacker ethics contain the rule "Mistrust authority–promote decentralization" (Fiery, 1994). This implies that governments should be mistrusted. Therefore, hacker ethics are in conflict with Hegel's doctrine. Hacker ethics also state "All information should be free." To let all information be free does not maintain the coherence of state (government) and is therefore wrong from the Hegelian viewpoint. Hence, Hegel's moral theory does not allow hacking in a general sense.

Self-Refutation

The theory of cultural relativism involves a contradiction (explicitly the reductio ad absurdum). It does not make any logical sense to claim that all moral judgements are relative, while maintaining that moral relativism itself is absolutely true (being non-relative). If all moral beliefs are relative, as relativists claim, absolutely true theories are an impossibility (Hare, 1986; Niiniluoto, 1991). This would also apply to cultural relativism.[3]

NON-DESCRIPTIVE USE OF ETHICS

The central task of moral philosophy is to determine what kinds of actions are right or wrong (Warburton, 1996; Hare, 1981). Descriptivistic theories such as cultural relativism fail to accomplish this mission: for example, cultural relativism does not explore what is morally right or wrong, but rather emphasizes what the moral habits of cultures are. It is argued that non-descriptivism offers more solid ground. Non-descriptive theories to discern what is truly morally right, what we ought to do, instead of appealing to our intuitions or cultural conceptions.

We see that many information security activities have strong moral dimensions. Information security protects against actions such as hacking or computer viruses, which may raise serious moral concerns. People who are victims of such activities (e.g., hacking or destructive viruses) are likely to express strong moral disapproval of the people responsible for such activities. However, without such personal experiences (Gattiker & Kelley, 1999), critical moral thinking or ethical education, etc., some users might feel neutral about such activities. In fact, computer ethics literature agrees that the ordinary computer user is often incapable of extending their moral reflection in cases where computers are involved. Several reasons to explain this problem are proposed. Moor (1985) explains this phenomena by the conceptual muddle and policy vacuum, i.e., the existing policies, such as legislation, do not cover computer ethics issues. Conger (et al., 1994) and Rubin (1994) argue that there is moral distance; e.g., the Internet creates a distance between users, and this distance decreases our moral sensitivity. Severson (1997) believes that people are in moral crisis. Siponen (2001) and Siponen and Vartiainen (2001) postulated that such users are conventionalists, i.e., they are pretty much incapable of engaging in critical moral thinking, but their acts reflect the prevailing moral views. As computer ethics issues are new and the existing prevailing moral conventions do not yet cover computing issues, conventionalists are unable to react to computer ethics issues. According to Dunlop & Kling (1992) and Rogerson (1996), computer users are under the spell of computers, forgetting the negative consequences of their usage. One may also argue, on the basis of Floridi (1999) and Gorniak (1996), that the existing ethical theories are inadequate to address computer ethics issues.

It is believed that some of these problems–such as moral crisis (Severson, 1997), moral distance (Rubin, 1994), conventional moral notions (Siponen, 2001; Siponen & Vartiainen, 2001) and the "spell" of computers (Dunlop & Kling, 1992; Rogerson, 1996) may be tackled with proper education. This effort may help in securing an organization's information systems, as well. However, in order that appealing to human morality would be useful for information security, the following conditions need to be met. An organization's business activities must be able to stand up to moral scrutiny and the organization's activities must not include double standards of morality. If the employees of the organization regard the organization's activities as improper, it is unlikely that they will respond to educational efforts. Organizations must display proper (moral) respect for their employees. If employers disrespect their employees, employees may have a reserved or negative attitude concerning such educational efforts. Organizations should facilitate an open climate for communication. Open environments for discourse as described by Habermas (1984, 1987) should be created. The immorality of acting against security policy can, with the help of ethical theories, be effectively argued. If an organization's employees are convinced that it is morally praiseworthy to follow the organization's security policy/procedures, they may be more willing to follow the policies/guidelines (consider the overriding thesis in Section 2). Equally, if an employer/educator is able to awaken employees' moral disapproval regarding acting against security policies/guidelines, it may be presumed that the employees will be more willing to follow security policies/guidelines (again consider the overriding thesis in Section 2).

In addition to the aforementioned prerequisites for using ethical education in organizations, the following guidelines for using ethics to persuade the listener can be used:

- Justify the principles (e.g., veil of ignorance/universality thesis as described below): State that the chosen principle is the best possible for the situation.
- Apply this principle and justify the results (justify the claim that the situation is morally acceptable and favourable).

These principles also facilitate the requirements of free will and autonomy. In other words, the problem of indoctrination might be avoided if the reasons for choosing certain ethical theories are justified. Indoctrination should be avoided since it halts autonomy–free will and autonomy are prerequisites for ethical decision-making (cf. Hare, 1964b, 1975).

Example of the Use of the "Veil of Ignorance"

Let us consider whether hacking is allowed, i.e., whether is it morally acceptable to obtain unauthorized access to information systems. This action is considered in the light of a simplified version of Rawls' theory of justice, which is affiliated with the universality principle proposed by Confucias (Singer, 1991), Kant, Hare (1963, 1981), Christian ethics (the Golden Rule), and Gewirth (1978). The limitations of the "veil of ignorance" are discussed in Kukathas and Pettit (1990) and Pogge (1989). The limits of universality theses are discussed generally in Siponen (2001), Siponen and Vartiainen (2001), and Hare's version of universalizability of moral judgement in Seanor and Fotion (1988).

Rawls (1971) proposes that the principles of justice–herein: should we allow hacking or not–should be selected under an imagined ignorance of our own role in the world (called a "veil of ignorance"). Under the veil of ignorance we are ignorant of our status, age, gender and the like. In doing this, the veil of ignorance strives to achieve impartiality since we are choosing principles that are equal for all; irrespective of our differences in terms of age, status, gender, colour of skin, cultural background–and systems, etc. This action prevents us from tailoring our moral principles to suit our role and disregarding the principles of the occupants of other roles (Hare, 1981; Rawls, 1972). Furthermore, the mentioned qualifiers (e.g., age, sex, reference to particular information systems) are ruled out since they are likely to be morally irrelevant to the choice of principles of justice (Rawls, 1971; Hare, 1963, 1981, 1989; Siponen, 2001). It just does not make any sense to claim that e.g., gender or ethnical background is relevant to the question of who is allowed to break into information systems.

So, in the case of hacking, given that the application of the veil of ignorance is desired, we need to imagine a situation in which we are unaware of our social status, age, sex, profession, etc. From behind this veil of ignorance we would need to ask ourselves whether we accept that hacking is allowed for *everyone*–i.e., anyone can break into any information systems at any time. This means that if we engage in hacking, we have to allow everyone else the same "right," even at the risk of them breaking into our systems. We submit that most of us, under the veil of ignorance, would not find hacking acceptable in a general sense. The reason being that if hacking is morally acceptable, there are no such things as company business secrets and individuals' privacy. Therefore "private" information such as medical information, social security numbers and financial information would be freely available for all to inspect. The result of which is most people feel more comfortable living in a society where hacking is not acceptable.

DISCUSSION, LIMITATIONS OF THE STUDY AND FUTURE RESEARCH QUESTIONS

The existing views on the relevance of ethics and human morality as a means of protection can be divided into two categories. First ethics and human morality can be used as a means of ensuring security and, second, the use of ethics is, at best, highly restricted, if not impossible. The underlying theories and justification for the two views differ. Proponents of the first view have not offered any theoretical background to justify their claims. In turn, the second view (the use of ethics is restricted to certain "security cultures") is based on cultural relativism (and hacker ethics) and can be classified as descriptivism. This paper explored the theoretical foundations of using human morality, criticised the descriptive view and proposed a non-descriptive approach for using human morality as a means of protection. Conceptual analysis was chosen as the research approach. This is a relevant choice since there is a lack of a solid ethical framework in current literature, and there is a need for a critical analysis of the existing descriptive view. Therefore, firm conceptual/theoretical foundations are needed as a first step, and conceptual analysis is the proper way to build such foundations.

Implications for Research and Practice

As for researchers and practitioners, this paper has clarified the point that human morality has a role in terms of security within organizations. This study also demonstrated that the argument against the relevance of human morality as a means of protection, on the basis of cultural relativism and hacker ethics, was fallacious. The study also clarified the limits of human morality (see Limits of the Findings). Furthermore this study has suggested several future research questions (see below). Moreover, with regard to practitioners, this paper has offered practical guidance on how ethics can be used as a means of ensuring security. It should be noted that the use of human morality at the fullest extent calls for educational skills and knowledge on moral philosophy. The application of ethical theories at an organizational level is not an impossibility. The question and challenge is where the educational and research/development efforts at universities and in organizations should be invested. Is an organization's security only a technical matter? Or has it strong social dimension, as is increasingly recognised by scholars?

Limitations of the Findings

The use of ethics as a means of ensuring security has a few limitations, however. First and foremost, it should be noted that the use of ethics is not a panacea. There is evil in the world (e.g., Warburton, 1996). This means that there are likely to be people who want to behave egoistically or maliciously, regardless of the moral status of such behaviour. Secondly, human morality can be used as a means of indoctrination. This may have short-term positive consequences from an organizational viewpoint. However this may turn against the organization in the long run. Moreover, intentional indoctrination is not morally acceptable since employee autonomy and free will are violated. Thirdly, moral philosophy would be useful for security only when an organization's security or business actions can withstand moral scrutiny. An organization's security or business activities are not per se morally right. For example, organisations may use information hiding (steganography) to carry out morally questionable activities. Fourthly, the use of ethics requires knowledge of ethical theories and persuasive discussion skills. Fifthly, without skillful education, the problem of indoctrination may come into play: A charismatic educator may ignorantly indoctrinate the employees. Sixthly, even if we want to behave morally right the fact remains that our decisions are deemed to be subjective. For example, it is possible, at least in theory, that after an application of the same ethical theory (e.g., the universality thesis) people would end up with differing views (philosophers are debating whether this is the case or not; see, e.g., Seanor & Fotion, 1988; Hare, 1999). The likelihood of the fourth problem may increase when we want to respect autonomy in moral matters, i.e., avoid indoctrination. Finally, we hold with R.M. Hare that, "Nobody can hope to write the last word on a philosophical subject; the most he can do is to advance the discussion of it by making at least some things clearer" (Hare, 1981).

Future Research Questions

There are a few research questions that future research should address. First, given that practitioners could use human morality as a means of ensuring security, studies that show how different ethical theories (non-descriptive) can be used for this purpose should be conducted. Secondly, empirical investigations are needed to explore whether computer users perceive morality as overriding over other nonmoral (such as financial) concerns (cf. Hare's overriding thesis in Section 2). Finally, further studies are needed to investigate what are effective persuasion strategies in ethical discussion (whilst at the same time avoiding indoctrination) and what are their effects. The possible research approaches would be conceptual analysis and empirical (theory testing/theory creating) research.

CONCLUSIONS

There is no doubt that the user has a significant role to play in building information security in organizations. Therefore, it is no wonder that the relevance of ethics as a means of ensuring security has been debated in information security literature. However, the current literature lacks solid theoretical foundations. Firm foundations are important to justify whether human morality has a role as a means of protection. It is necessary to examine the question of applicability and how to appeal to human morality as a means of protection. To build this foundation, we look to the wisdom of ethical theories. To increase our understanding of the fundamental differences between ethical theories, the theories were classified into two categories, descriptivism and non-descriptivism. It was shown that the question of whether ethics and human morality have any relevance can be retraced to Hare's overriding thesis (moral overrides all other concerns). The descriptive claim that ethics cannot serve as a means of ensuring security, based on cultural relativism, was considered. It was shown that this descriptive view–that ethics is of little use in securing organizations–encompasses several problems. To avoid these problems, an alternative non-descriptive approach was proposed and an example of its use was given. Four prerequisites and a two-step persuasion guideline for using ethics as a means of ensuring security were put forth. Moreover, limits of the use of human morality were considered. Finally, several future research questions were suggested.

ACKNOWLEDGEMENTS

I am grateful to Mr. Pekka Abrahamsson, Prof. Juhani Iivari, Prof. Marius Janson, Prof. Jussipekka Leiwo and Dr. Kari Väyrynen for their comments on the earlier version of this paper. I would also like to thank the anonymous reviewers of this special issue for their comments.

ENDNOTES

1 One may raise an objection towards my use of the word "culture" by stating that an association, in this example, is not a culture, but it exists within a culture. If tribes or hackers are considered to be "cultures" then associations should be accorded the same treatment. At any rate, further considerations concerning whether an association is a culture or not are irrelevant for this example. The example given can be safely modified so that the association is not a culture, but is within a culture, and the values of this culture are exactly the values of the association (and the company is outside of this culture).

2 Hegel can perhaps be seen to be a kind of "relativist" as, in his view, we must recognise the fact that we all have a history. Because of this, Hegel believes that we cannot share the same categorical imperatives, as put forth by Kant, for instance. This may violate Hume's law provided that the different histories are regarded as "is" matters from which an "ought" conclusion is inferred.

3 Similar reductio ad absurdum fallacies, with respect to cultural relativism, have been formulated. Hare (1986), for example, has pointed out another such fallacy he claims that the existence of such fallacies raises objections to all descriptive views, given that they lead to cultural relativism. The association example provided earlier illustrates another contradiction. The association accepts hacking while the company does not, and the "culture" of the company is just a short distance away from "the culture" of the association. Hence, there seems to be a contradiction in saying that the exact same action in all respects (e.g., hacking) is acceptable (by the association) and not acceptable (by the company). This idea also conflicts with the supervenience relation, e.g., see (Hare, 1952, 1984; Kim, 1984, 1991), as it is inconsistent to claim that the same action can be simultaneously both wrong and right, depending on the culture.

REFERENCES

Anderson, T. E. (1993). Management guidelines for PC security. *Proceedings of the 1992 ACM/SIGAPP Symposium on Applied Computing (vol. II): Technological Challenges of the 1990's*. Kansas City, USA.

Baskerville, R. (1995). The second-order security dilemma. In Orlikowski, W., Walsham, G., Jones, M. and DeGross, J. (Eds.), *Information Technology and Changes in Organizational Work*, 239-249. London: Chapman & Hall.

Conger, S., Loch, K. D. and Helft, B. L. (1994). Information technology and ethics: An exploratory factor analysis. *Ethics in the Computer Age Conference Proceedings*, Gatlinburg, Tennessee, November 11-13.

Dancy, J. (1994). Why there is really no such things as the theory of motivation. *Proceedings of the Aristotelian Society.*

Dhillon, G. and Backhouse, J. (2000). Information system security management in the new millennium. *Communications of the ACM*, 43(7), 125-128.

Dunlop, C. and Kling, R. (Eds.). (1992). Social relationships in electronic commerce–Introduction. In *Computerization and Controversy–Value Conflicts and Social change*, 322-329. New York: Academic Press.

Fiery, D. (1994). Secrets of a super hacker. *Loompanics Unlimited*. Washington, DC: Port Townsend.

Floridi, L. (1999). Information ethics: On the philosophical foundation of computer ethics. *Ethics and Information Technology*, 1(1), 37-56.

Gattiker, U. E. and Kelly, H. (1999). Morality and computers: Attitudes and differences in moral judgements. *Information Systems Research*, 10(3), 233-254.

Gewirth, A. (1973-1974). The "is/ought" problem resolved. *Proceedings and Addresses of the APA*, 47, 34-61.

Gewirth, A. (1978). *Reason and Morality*. Chicago, IL: The University of Chicago Press.

Gewirth, A.. (1982). *Human Rights: Essays on Justification and Applications*. Chicago, IL: The University of Chicago Press.

Gewirth, A. (1996). *The Community of Rights*. Chicago, IL: The University of Chicago Press.

Goodhue, D. L. and Straub, D. W. (1989). Security concerns of system users: A proposed study of user perceptions of the adequacy of security measures. *Proceedings of the 21nd Hawaii International Conference on System Science (HICSS)*.

Gorniak, K. (1996). The computer revolution and the problem of global ethics. *Science and Engineering Ethics*, 2(2).

Habermas, J. (1984). *The Theory of Communicative Action–Reason and the Rationalisation of Society,* I, Boston, MA: Beacon Press.

Habermas, J. (1987). *The Theory of Communicative Action–The Critique of Functionalist Reason*, Vol II, Beacon Press, Boston, MA.

Hare, R. M. (1952). *The Language of Morals*. Oxford, UK: Oxford University Press.

Hare, R. M. (1963). *Freedom and Reason*. Oxford, UK: Oxford University Press.

Hare, R. M. (1964a). *The Promising Game*. Revue Internationale de philosophie 70.

Hare, R. M. (1964b). Adolescents into adults. In Hollins, T. C. B. (Ed.), *Aims in Education*. Manchester.

Hare, R. M.. (1975). Autonomy as an educational idea. In Brown, S. C. (Ed.), *Philosophers Discuss Education*. Macmillan.

Hare, R. M. (1976). Some confusions about subjectivity. In Bricke, J. (Ed.), *Freedom and Morality*. Kansas University Press.

Hare, R. M. (1981). *Moral Thinking: Its Levels, Methods and Point*. Oxford, UK: Oxford University Press.

Hare, R. M. (Ed.). (1984). Supervienience. *Proceedings of Aristotelian Society*, suppl. 58. Reprinted In (1989), *Essays in Ethical Theory*. Oxford, UK: Clarendon Press.

Hare, R. M. (1986). A reduction ad absurdum of descriptivism. Shanker, S. (Ed.), *Philosophy in Britan Today*. London, UK: Croom Helm.

Hare, R. M. (Ed.). (1989). Principles. In *Essays in Ethical Theory*, 48-65. Oxford< UK: Oxford University Press.

Hare, R. M. (Ed.). (1997). A taxonomy of ethical theories. In *Sorting out Ethics*. Oxford, UK: Oxford University Press.

Hare, R. M. (1999). *Objective Prescriptions and Other Essays*. Oxford, UK: Oxford University Press.

Harrington, S. J. (1996). The effect of codes of ethics and personal denial of responsibility on computer abuse judgements and intentions. *MIS Quartely*, September, 20(3).

Järvinen, P. (1997). The new classification of research approaches. Zemanek, H. (Ed.), *The IFIP Pink Summary–36 years of IFIP*, Laxenburg, IFIP.

Kant, I. (1993). *The Moral Law: Groundwork of the Metaphysic of Morals*, Routledg, London.

Kim, J. (Ed.). (1984). Concepts of supervenience. *Philosophy and Phenomenological Research*, 45, 153-176. Reprinted in (1993), *Supervenience and Mind*. Cambridge University Press.

Kim, J. (Ed.). (1991). Supervenience as a philosophical concept. *Metaphilosophy*, 21, 1-27. Reprinted in (1993), *Supervenience and Mind*. Cambridge University Press.

Kowalski, S. (1990). Computer ethics and computer abuse: A longitudinal study of Swedish University students. *IFIP TC11 6th International Conference on Information Systems Security*.

Kukathas, C. and Pettit, P. (1990). *Rawls–A Theory of Justice and its Critics*. California: Stanford University Press.

Ladd, J. (1982). Collective and individual moral responsibility in engineering: Some questions. *IEEE Technology and Society Magazine*, 1(2), 3-10.

Ladd, J. (1989). Computers and moral responsibility: A framework for an ethical analysis. In Gould, C. (Ed.), *The Information Web: ethical and Social Implications of Computer Networking,* 207-227.

Leiwo, J. and Heikkuri, S. (1998a). An analysis of ethics as foundation of information security in distributed systems. *Proceedings of the 31st Hawaiian International Conference on System Sciences (HICSS-31).*

Leiwo, J. and Heikkuri, S. (1998b). A group-enhanced ISSI model for secure interconnection of information systems. *Proceedings of the IFIP TC11, 14th International Conference on Information Security (IFIP/Sec'98).*

Loch, K. D. and Carr, H. H. (1991). Threats to information system security: An organizational perspective. In *Proceedings of the Twenty-Fourth Annual Hawaii International Conference on System Sciences (HICSS).*

MacIntyre, A. (1981), After virtue. *A Study in Moral Theory*. London, UK.

Mackie J. L. (1981). *Ethics, Inventing Right and Wrong*. London: Penguin Group

Mautner, T. (Ed.), (1996). *A Dictionary of Philosophy*. Oxford, UK: Blackwell.

McLean, K. (1992). Information security awareness–selling the cause. *Proceedings of the IFIP TC11 (Sec'92).*

Moore, G. E. (1903). *Principia Ethicia*. UK: Cambridge.

Neumann, P. G. (1999). Inside risks: risks of insiders. *Communications of the ACM*, 42(12), 160.

Niiniluoto, I. (1991). What's wrong with relativism. *Science Studies*, 4(2), 17-24.

Outga, G. (1972). *Agape: An Ethical Analysis*. Yale University Press.

Parker, D. B. (1998). *Fighting Computer Crime–A New Framework for Protecting Information*. USA: Wiley Computer Publishing.

Perry, W. E. (1985). *Management Strategies for Computer Security*. Boston, MA: Butterworth Publisher.

Pogge, T. W. (1989). *Realizing Rawls*. Cornell University Press.

Popper, K. (1948). What can Logic do for Philosophy? *Aristotelian Society, Supplementary*, 22.

Rawls, J. A. (1972). *A Theory of Justice*. Oxford, UK: Oxford University Press.

Rogerson, S. (1996). The ethics of computing: The first and second generation. *The Business Ethics Network News*, 6.

Rubin, R. (1994). Moral distancing and the use of information technologies: the seven temptations. *Proceedings of Ethics in the Computer Age Conference*, Gatlinburg, Tennessee, November 11-13.

Sabine, G. H. (1963). *A Gistory of Political Theory*. Third edition. London, UK.

Sandel, M. (1982). *Liberalism and the Limits of Justice*. UK: Cambridge University Press.

Seanor, D. and Fotion, N. (Eds.). (1988). The levels, method and points. In *Hare and Critics–Essays on Moral Thinking*, 3-8. Oxford, UK: Oxford University Press.

Searle, J. (1964). *How to Derive "Ought" From "Is."* Ph. Rev. 73.

Severson, R. J. (1997). *The Principles of Information Ethics*. Armonk (N.Y.) M. E. Sharpe cop. USA.

Singer, P. (1991). *A Companion to Ethics*. UK: Blackwell.

Siponen, M. T. (2000). A conceptual foundation for organizational information security awareness. *Information Management & Computer Security*. 8(1), 31-41.

Siponen, M. T. (2001). The relevance of software rights: An anthology of the divergence of sociopolitical doctrines. *AI & Society*, 15(1-2), 128-148.

Siponen, M. T. and Vartiainen, T. (2001). End-user ethics teaching: Issues and a solution based on universalization. *Proceedings of the 34th Hawaii International Conference on Systems Sciences*.

Smart, N. (1986). Relativism in ethics. In Macquarrie, J. and Childress, J. (Eds.), *A New Dictionary of Christian Ethics*. London, UK: SCM Press LTD.

Smith, M. (1984). *The Moral Problem*. Oxford, UK: Blackwell.

Straub, D. W. (1990). Effective IS security: An empirical study. *Information System Research*, June, 1(2), 255- 277.

Straub, D. W. and Nance, W. D. (1990). Discovering and disciplining computer abuse in organization: A field study. *MIS Quartely*, March, 14(1).

Straub, D. W. and Widom, C. P. (1984). Deviancy by bits and bytes. In Finch, J. H and Dougall, E. G. (Eds.), *Computer Security: A Global Challenge, Proceedings of the Second IFIP International Conference on Computer Security (IFIP/Sec '84)*.

Taylor, P. W. (1975). *Principles of Ethics–An Introduction*. Encino, CA: Dickenson Publishing.

Thomas, R. K. and Sandhu R. S. (1994). Conceptual foundations for a model of task-based authorizations. *Proceedings of the 7th IEEE Computer Security Foundations Workshop*.

Thomson, M. E. and von Solms, R. (1998). Information security awareness: Educating our users effectively. *Information Management & Computer Security*, 6(4), 167-173.

Vardi, Y. and Wiener, Y. (1996). Misbehavior in organizations: A motivational framework. *Organization Science*, March-April, 7(2), 151-165.

Warburton, N. (1996). *Philosophy: The Basics*. Second Edition. Cornwall, UK: T J Press Ltd Padstow.

Chapter 20

The Government "Downunder" Attempts To Censor The Net

Geoffrey A. Sandy
Victoria University, Australia

In western tradition information ethics has its origins in Athenian democracy. It was characterised by an oral culture and freedom of speech. Later, after a great struggle, freedom of written expression was added. In this age of electronic networks freedom of access to the Internet must be added. Currently, this freedom is under sustained attack worldwide. The Australian Government has joined this attack with the passage of the "Broadcasting Services Amendment (Online Services) Bill 1999. The legislative purpose is to regulate access to content that is offensive to a "reasonable adult" and unsuitable for children. This chapter reports the results of an analysis of the primary sources regarding the Bill. Specifically, it reports on the important issues that were addressed in the parliamentary hearings and debates. It also comments on the success of the legislation after 8 months of operation. Documentation of the Australian experience should inform other countries that are currently attempting to understand and resolve these complex issues, or for those who will attempt to do so in the future.

BACKGROUND

Australians claim they live in an open society that guarantees freedom of expression and, by world standards they do. It is however, a society that lacks a Bill of Rights and has no equivalent of the American first amendment. It has a long history of censorship offline that predates the establishment of a federal system of government in 1901. Censorship legislation exists at both the federal and the state/

Previously Published in *Managing Information Technology in a Global Economy*, edited by Mehdi Khosrow-Pour, Copyright © 2001, Idea Group Publishing.

territory levels. It differs from state to state. Recently, the federal and state governments turned their attention to the Internet and passed legislation to censor selected material. This legislation also differs from state to state.

On 1 January 2000, the federal government's *Broadcasting Services Amendment (Online Services) Bill 1999* (hereafter the Bill) came into force. The purpose of the Bill is to establish a framework for the regulation of the content of online services. It is this legislation that is the subject of this paper. The Bill seeks to:

1. provide a means of addressing complaints about certain Internet content.
2. restrict access to certain Internet content that is likely to cause offence to a reasonable adult.
3. protect children from exposure to Internet content that is unsuitable for them.

On receiving a complaint about Internet material the Australian Broadcasting Authority (ABA) can require an Internet Service Provider (ISP) to take down X rated or refused classification material (RC) hosted onshore, and, to take all reasonable steps to prevent access to X rated or RC material hosted offshore. In respect to restricted (R) classified onshore content the ABA must be satisfied that restricted access arrangements are in place. No proposal is made in respect to offshore R classified material. The Bill expressly specifies time frames for the take down process and penalties for non-compliance. The Bill expressly does not apply to live Internet content such as news groups, chat channels, or e-mail. A body to monitor online material envisaged by the Bill has been established.

Australia uses a classification system to determine the degree of censorship that is to be applied to a variety of offline media that include films, videotapes, publications, and video games. The Classification Board of the Office of Film and Literature Classification (OLFC) undertake the classification. The state and territory governments also use this. Classification decisions apply criteria that are part of the National Classification Code. The most important "test" that determines a particular classification is the concept of the "reasonable adult." Such a person is defined officially as "possessing common sense and an open mind, and able to balance opinion with generally accepted community standards" (OLFC, 1999b).

RESEARCH QUESTIONS

The federal government's decision to censor the Internet demands an understanding of, and resolution of, many complex issues. These issues have relevance for government in most countries. These issues are not necessarily new and the experience of other countries in addressing them may differ from Australia. For some countries these remain to be addressed. Documentation of the recent Australian experience in censoring the Internet should inform other countries that are attempting to understand and resolve these complex issues or those who will attempt to do so in the future. Space does not permit detailed reference to other countries. The Australian experience should particularly inform those countries that share its relative openness as a society and its high usage of Internet technology. The primary research question is - *What are the major issues associated with the Australian federal government's decision to censor the Internet?* A series of secondary questions follow:

1. Is government censorship of Internet pornography justified?
2. Is Internet pornography a major concern of the Australian community?
3. Is the Bill enforceable?
4. Will the Bill protect minors from unsuitable material?
5. Will the Bill protect adults form offensive material?
6. Will the Bill seriously inhibit the development of the Australian Internet industry?
7. Is the Bill a result of the influence of the "conservative right?"

Questions 1 to 6 should be relevant to most countries that have a high level of Internet usage. Question 7 may have less relevance to other countries.

RESEARCH METHOD

In order to address these research questions a literature review and a detailed analysis of primary source documents was undertaken. The literature review surveyed censorship in Australia, and to a lesser extent, other countries. It mainly concentrated on the history and current arrangements of Australian government regulation of offline media, together with reports and studies concerning online regulation, that pre-date the Bill

The primary source documents are:

1. The Official Committee Hansard of the hearings from the submitters on the *Broadcasting Services Amendment (Online Services) Bill 1999* by the Senate Select Committee on Information Technologies.
2. Senate Hansard of the second reading of the Bill.
3. House Hansard of the second reading of the Bill.
4. The *Broadcasting Services Amendment (Online Services) Bill 1999.*

Throughout the paper reference is made to (1) above as the "hearings," (2) and (3) as the "debates" and (4) as "the Bill." The abbreviated in-text citation used throughout for (1) is "SSC" for Senate Select Committee), (2) is "SH" Senate Hansard and (3) "HH" House Hansard. Full details are provided in the references section.

The Bill was prepared by the ruling conservative coalition but was opposed in part or whole by the opposition parties. During the parliamentary debate on the Bill the balance of power in the Senate rested with two Independents who needed to vote with the Coalition for the Bill to be passed.

THE JUSTIFICATION FOR GOVERNMENT CENSORSHIP OF INTERNET CONTENT

In a relatively open society like Australia all political parties support the principle that adults should be able to read, hear and see what they want. None would consider this as an absolute right to "freedom of expression. All would accept some qualification to this freedom. It is how this is interpreted that gives rise to different views about the justifiable degree of government censorship

The classic libertarian position (Whittle, 1998; Whitaker, 1994; Carol, 1994) is to treat freedom of speech as indispensable to the open society. If the government wishes to censor some type of speech, like pornography, then the onus of proof of demonstrable harm rests with the government. State intervention to censor pornography without empirically proven evidence of social harm is an illegitimate exercise of state power in an open society. During the hearings and the debates offensive and unsuitable material was almost invariable equated with pornography.

Most research studies on pornographic harm are for media other than the Internet. Australia lacks empirical research in this area. As is common for any contentious area each group holding different views claim the research findings support their particular view. Pornographic harm is no exception. The findings summarised here (Felson, 1999; Segal, 1992; Wilson and Nugent, 1987) are reviews that may be described as "independent". However, the reviews by Graham (2000) and Carol (1994), although they espouse the "libertarian" view, are summarised because they are well argued and comprehensive.

The findings are that:

1. A causative link between sexually explicit material with no violence and:
   - acts of violence (non consensual acts) has not been established.
   - change in male attitudes towards women (for the worse) has not been established.
2. In regard to sexually explicit material with violence:
   - a causative link between exposure to sexually violent material and sexually aggressive behaviour (towards woman) has not been proven.
   - the causes of the impulses of "child abusers" and rapists are found in childhood experiences, especially highly sexually repressive backgrounds often stemming from religious beliefs in their families that stigmatised all sexual responsiveness and expression.
   - the significant predictors of rape in an area is population size, proportion of young adults and percentage of divorced couples – not pornography.

3. In regard to violent material with nothing sexually explicit:
    - no conclusive causative link between media violence and violent offences is established.
    - notwithstanding the point above, many researchers are convinced that there is sufficient tentative evidence of harmful effects to warrant caution.

In regard to minors and pornographic harm there are no reported studies for obvious ethical reasons. In the case of images of paedophilia or images of incest with minors it is widely agreed that these be illegal because it is assumed that production of such images involve non-consensual acts.

There is little empirical support to justify censorship of pornography in an open society on the grounds of being harmful to adults. During the hearings and parliamentary debates the "free speech" issue was raised (SH 5199; 5207; 5213). The Democrats specifically argued that "blocking lists used in filters should be fully disclosed" and was concerned about "inappropriate and inadvertent blocking" (SH 5201). Supporters of the legislation argued that it did not suppress free speech (SCC150; HH 6907; SH 5219). Little beyond assertions was said about pornographic harm to adults.

THE NEED FOR GOVERNMENT CENSORSHIP OF INTERNET PORNOGRAPHY

The coalition claimed that the Bill was a response to a major concern expressed by the Australian community about Internet pornography (SH 5218; HH 6907). They argued that the Bill was needed to regulate online what was already regulated offline. During the debates the coalition claimed that the Bill had attracted little or no criticism from the public (SH 5218). Reference was made by Minister Alston, who had responsibility for the Bill, to an unnamed American study of parent need that was claimed to be relevant for Australia. He then asserted that Internet pornography is a major social problem for Australia (SH 5218). Indeed, the Coalition members believed that the Internet is awash with pornography with a claim that 60% of Internet content is pornography (SSC 49; 73). Internet Industry representatives at the hearings reacted with incredulity to this claim (SSC 49; 73).

No comprehensive study on community attitudes to Internet pornography has been conducted in Australia. Both opponents and supporters of the Bill provided anecdotal evidence to support their case (SH 5210; 5212). What empirical evidence exists suggests that Internet pornography is not a major community concern (SSC 260ff; 297-299). The spokesperson for the Australian Library and Information Association (ALIA) was questioned at the hearings about the measures in place in libraries throughout the country to protect minors from accessing unsuitable Internet material. Coalition members expressed disbelief and became hostile in their questioning after she claimed that children accessing inappropriate material had not been raised as an issue with the Association (SSC 190ff; 103).

Both opposition members (SH 5199; 5210) and submitters to the hearings (SSC 48-49; 70) pointed out that the Bill was not needed because the existing laws, especially, the Crimes Act, are adequate to deal with illegal activity online. An industry spokesperson at the hearings succinctly put it "as far as we know there are no illegal activities that become less illegal on the Internet" (SSC 209). However, the hearings were told that some uncertainty exists as to whether the Crimes Act does cover the Internet (SCC 123). The coalition indicated it would legislate to rid the Act of any uncertainty (HH 6907).

Supporters of the Bill indicated a concern that the purveyors of over 90% of pornography into Australia was from overseas, and that there is a need to protect Australians from "foreign filth" alien to Australian culture (HH 6916ff). The independent Senator Harradine, a strong supporter of the Bill, asserted that the US Supreme Court's interpretation of the right to freedom of expression in relation to " The Communications Decency Act" is not relevant to Australia (SSC 70). The coalition was sensitive to parallels being drawn between the Bill and the failed "Communications Decency Act" (HH 6920). A strong parliamentary opponent of the Bill lamented the violence portrayed in American film and television that dominate in Australia, and the lack of willpower by the government to regulate a phenomenon likely to be more harmful than sexual explicitness (SH 5213).

THE ENFORCEABILITY OF THE BILL

A major criticism of the Bill is that it is unenforceable (SSC 163ff; 35-36; SH 5136). A spokesperson from the Internet industry summarised the view of many when he stated that "the universal attitude to this proposal is sheer disbelief that any government would attempt to introduce something like this with any expectation that it might be successful" (SSC 113). The four major issues associated with enforceability was the:

1. nature of the Internet vis-à-vis other media
2. decision to make ISPs and backbone providers responsible for blocking Internet content
3. effectiveness of software products in blocking Internet content
4. dominance of overseas hosted content

A common charge made against the coalition is that the Bill is based on a mistaken view about the nature of the Internet. Specifically, the supporters of the Bill identified broadcasting (narrowcasting) as the closest media analogy to the Internet (SCC 119; 132ff), and for classification purposes it be treated as analogous to film and video (SSC 71). When the representative from the ABA was asked at the hearings whether he believed it was able to administer and enforce the legislation the answer was in the affirmative (SSC 31). Opponents of the Bill argued that the Internet is unique and approaches to the traditional media do not apply (SSC 208ff; 47ff; SH5205ff) or at least it is more analogous to books (SSC 65ff; SH 5199) or the telephone (SSC 67) than film.

Those opposed to the Bill argued that the best place to block content is at the receiver and not the ISP or backbone provider (SSC 151ff; 243ff). The findings of a report by the Commonwealth Scientific and Industrial Research Organisation (CSIRO 1999) were given prominence in the hearings and debates. It was treated as authoritative by both sides. The report considered blocking access at the ISP or backbone provider level to be largely ineffective. Instead it recommended the use of filtering software by Internet users. It evaluated the effectiveness of filter products and commented on their deficiencies.

The coalition acknowledged the deficiencies of the filter products but looked forward to a future where technology had developed to such a state that it would make the Bill more enforceable (HH 7978). Until then the coalition responded by stating "the fact that some aspects of the bill are in the opinion of some in the industry, unlikely to succeed is not argument enough, for the government to ignore this issue and to allow the Internet to go unchecked as a conduit to the world's unsavoury material for our children" (HH 7978).

The government was reminded by industry spokespersons at the hearings (SCC 63ff; 127ff) and in the debates (HH 7970; SH 5136) that the legislation will not work because over 90 per cent of the content is hosted offshore. Supporters of the Bill were also reminded of the ease with which domestic sites could move offshore to escape enforcement (HH 6913).

THE EFFECTIVENESS OF THE BILL – PROTECTION OF MINORS

One of the aims of the Bill is to protect minors, that is children under 18 years of age, from Internet material that is unsuitable for them. The protection of children from pornographic harm dominated the hearings and the debates. Naturally, there was unanimous support for this sentiment although the coalition did attempt to portray the opposition members as "soft" on porn and, unwilling to support measures that protect children from it. Again, there is unanimity that parents or guardians have primary responsibility to regulate their child's access to Internet content with support from the government in empowering them with education and technology. The coalition stressed the importance of the Bill because "we have a unique opportunity to set a standard for the world rather than step away from our responsibilities. In effect we are taking an international leadership position" (HH 7977).

The major differences between those supporting the legislation and those opposed centred on the:

1. support for computer illiterate parents.
2. support for children of irresponsible parents.
3. trustworthiness of the technology in protecting minors.
4. preservation of the Internet as an adult medium.

The coalition members argued that the Bill would support parents who lack the technical knowledge necessary to properly supervise their children's access to the Internet (SSC 51; SH 5219; HH 7977). Much was made by them of portraying children as more technically competent than their parents, and because of this, are able to deceive their parents about their Internet use. Implicit in the arguments of some supporters of the Bill was that it would counter irresponsible parents who did not properly supervise their children's Internet access (SSC 51; SH 5219; 5212). The Bill was considered necessary to protect children from their own (irresponsible) parents.

Coalition members at the hearings adopted a similar view towards library staff in regards to supervision of Internet access by children in libraries. They suggested that library staff are too busy to be monitor breaches of policy, and given the tenor of questioning an implicit suggestion that the library policy on this matter borders on the irresponsible (SSC 192ff; 103).

A common argument made by those opposed to the Bill was that not only is the blocking software ineffective in protecting minors but the charge that the government is deceiving parents into believing the technology is trustworthy (HH 7970ff; SSC 257-258; 104). Thus, parents are "lulled into a false sense of security" (SSC 245) and therefore "the Bill could do more harm than good" (SSC 63). Further, it was argued that the government by putting its faith in software that does not work results in a lost opportunity to educate and empower parents to take responsibility for supervision of the Internet by their children (SSC 6914).

The representative of the Australian Computer Society at the hearings spoke for many when he argued that the Internet should remain an adult medium and should not be transformed into a media suitable only for children (SSC 72). This he claimed would be the effect of the legislation. The coalition countered that the Internet warranted no special treatment in regards to government regulation compared to the other media (SH 5219; HH 6970).

THE EFFECTIVENESS OF THE BILL – PROTECTION OF ADULTS

The main discourse at the hearings and in the debates was about the effectiveness of the Bill in affording protection to children. However, the legislation also aims to "restrict access to certain Internet content that is likely to cause offence to a reasonable adult." It is surprising that the opponents of the Bill did not criticise it for not including the qualifier "unsolicited" material, otherwise it amounts to the government protecting adults from their own actions.

Even more surprising was that little was said about equating harm of the "reasonable adult" with feelings of being offended by an image and/or text. This is central to the governments system of censorship of any media. Vital to a view on pornographic harm is the meaning and power of the pornographic image (Sandy, 1999). The literature almost exclusively discusses harm (to woman) from pornography produced and consumed by men. It ignores, for instance, pornography consumed by woman, produced and consumed by woman, and, gay pornography. One view (Cowie, 1992; Masson, 1990; Morgan, 1980) is that the pornographic image is a master text to practices it portrays. It does represent real (bad) attitudes and desires, and, it teaches the consumer these attitudes, and to expect to recreate these practices in reality. As indicated previously there is little empirical support for this view. The opposing view (Gibson and Gibson, 1993; Cowie, 1992; Segal, 1992) is that the pornographic image is a signifying system and a fantasy scenario. What is portrayed is not the object of desire but a scenario in which certain wishes are presented. This may involve scenarios that are illegal as acts. Fantasy is clearly a separate realm from reality and cannot be taken at face value.

Few opponents of the Bill reiterated the classic libertarian argument as it applies to contemporary Australian society. A society composed of many communities of widely differing standards in relation to sexuality, gender relations and sexual expression. One of Australia's greatest assets is a social structure that is highly pluralistic. In reality, the "test" of the "reasonable adult" results in the majoritian norm being forced on all communities. In an open society like Australia, speech that is deemed illegal or restricted should depend on "proving" its demonstrable harm. In the absence of this mere offensiveness to a group is no justification for censorship by the state

THE IMPACT ON THE DEVELOPMENT OF THE INTERNET INDUSTRY

A number of issues related to the impact of the Bill on the nascent Internet e-commerce market were raised. As one opponent of the Bill put it "if you wanted to hurt the Internet and you wanted to hurt Internet electronic commerce, this would be one of the best ways to start" (SH 5211).

First, concern was expressed that given that the United States is Australia's primary Internet trading partner, efficient online trading would need to have compatible and consistent legal structures with that country. This would not be the case if the Bill passed (SSC 208). Second, concern was expressed that compliance costs imposed on ISPs, especially those in regional Australia may retard development, especially the small ISPs (SSC 151; 208). Third, it was argued that the current situation of a relatively unregulated Internet gave Australia a competitive advantage. This would disappear with the passage of the Bill (SSC 208).

Fourth, Australian content creators and distributors would be forced to go underground or offshore to avoid the compliance costs (SSC 103; SH 5198; HH 7975). Fifth, concern was expressed that overseas investors would get the wrong impression that the Australian government appears to focussing on pornography rather than more substantive issues like e-business penetration, security, privacy etc. (SSC 79; 90; SH 5210). Finally, the CSIRO Report (1998) was cited in support of the argument that implementation of blocking technologies would degrade the Internet (SCC 75).

The coalition did not accept these arguments. Indeed as one coalition member stated "they cannot in one breadth claim that the Internet is going to come to a grinding halt and in the next breadth say that the Net is so powerful, so above anything governments within jurisdictions can do to improve the quality of the content it carries, that it will just ride over the top. You cannot have it both ways" (HH 6920).

OPPORTUNISTIC POLITICS OF THE BILL

The coalition and other supporters of the Bill argued that the legislation was in response to community concerns about Internet pornography. Principally, it was needed to protect children from "a litany of filth and cyber-septic" material (HH 6916ff). The government assured all that the Bill did not single out the Internet for "special treatment" but rather it aimed to regulate selected content online in the same way as it did offline.

Those opposed to the Bill asked that if the Bill is not wanted by the community, largely unenforceable, largely ineffective in protecting minors and adults, and is likely to have an undesirable impact on the growth of the new economy what then is the government's motive for it? For them (SH 5135ff; HH 6911ff; 7971ff) the most obvious explanation the Bill was brought to parliament at this time (June1999), and with unseemly haste, was to appease the conservative independent senate member (Harradine). He had the decisive vote on government legislation in a senate because of equal numbers of the ruling coalition and opposition parties. The other independent was very sick and preoccupied at facing criminal charges. In particular the coalition needed Harradine's vote to pass the privatisation of Telstra legislation. The reward was the promise to allocate a substantial part of the sale proceeds to his home state of Tasmania and to "get tough" on Internet pornography. The senator has a long history of crusading action against contraception, abortion, sexual explicitness and violence in society (Marr, 1999). During the debates he lamented the fact that the Bill did not make illegal the X and R classified material (SH 5215ff). Censorship of the Internet was dear to his heart.

THE CONSERVATIVE RIGHT AND THE BILL

How a society treats "pornography" is a good test of its "openness." Pornography is subversive. It challenges fundamental values and names deepest fears. It challenges the institutions of the "traditional" family and church. It exposures the hypocrisy on gender relations, class bias and sexuality. It tells us things about ourselves that we do not wish to acknowledge. It is not surprising that many will demand that this speech be suppressed.

Whilst acknowledging that winning the support of Harradine was a factor in explaining the introduction of the legislation at this time it is an incomplete explanation. What is argued here is that the Bill is primarily due to the influence of the conservative right, especially the religious right. Censorship of the Internet is yet another example of the persuasive influence of this group in Australian society. These people wish to compel others to accept their views on gender relations, sexuality and sexual expression. The prevailing conservative discourse in Australia provides fertile ground for accommodation of these views.

The most obvious mechanism for this influence on the government is a group of religiously conservative coalition members known as the Lyons Forum (EFA, 1998). The Electronic Frontiers Australia (EFA) an organization dedicated to an uncensored Internet has confirmed that 33 coalition members of parliament belong to the Forum. It is suspected the number is much higher. No opposition members belong to it. As could be expected the Lyons Forum is well represented on the Senate Select Committee on Information Technologies. The Lyons Forum, together with Harradine and other fellow-travellers have been very influential in legislation that severely regulated "phone sex," video games, and more recently the failure to support the change from X rated to "non-violent erotica" videos (Graham, 2000; Marr, 1999). Harradine has been a member of the Senate Select Committee for many years.

There is evidence that suggests that the conservative right is not only or even primarily concerned with pornography. The Bill treats Internet content as a film or video for classification purposes. EFA point out that for 1997–98, that 68% of films and 71% of videos were classified R for *"other" reasons*, that is, adult themes. A minority were classified based on "sex" or "violence." It believes that the legislation is about a great deal more than pornography. Adult themes include "verbal references to and depictions associated with issues such as suicide, crime, corruption, marital problems, emotional trauma, drug and alcohol dependency, death and serious illness, racism, religious issues" (EFA, 1999a). EFA argue that by "legislating that adult discourses be placed behind locked gates, hidden from search engines, is a sure means of discouraging people from becoming informed" (EFA, 1999a). They view the legislation as a "full-frontal" attack of freedom of speech. In Popper's terms (1969, 1973) the conservative right may be considered as enemies of the open society, and they would appear to be alive and well "downunder."

CONCLUSION

The primary ethical question relevant here is - *Under what circumstances is government censorship of any media justified?* In an open society the onus is on those who wish the state to censor to "prove" demonstrable harm. Worldwide empirical evidence over many decades does not support a causal link between pornographic images and harmful actions done to other persons. Freedom of speech should still prevail even if the majority believe that speech to be offensive. The evidence in Australia suggests that the vast majority do not view Internet pornography as a major social problem. The risk of a reasonable adult inadvertently accessing material that may give offence does not justify the state making it illegal or mandating restrictions.

Protection of minors from accessing unsuitable material rests primarily with the parents. They can choose to use blocking technology or "white list" services. It should be made very clear to parents that such technology is very inaccurate. A better strategy is for parents and schools to provide a sound moral education for the child. An education that teaches positive attitudes to sexuality and gender relations. One that combats sexism, misogyny and homophobia. Then children are better equipped to deal with inadvertent access of unsuitable material. It is this very strategy that the conservative right oppose, especially within the school system.

POSTSCRIPT AND FURTHER RESEARCH

On the 30 June 2001 the Online Services Bill will have operated for 18 months. It will be opportune to investigate whether or not this approach to censoring the Internet has been successful. However, it may be several years before the full impacts of the Bill can be fully assessed.

After 8 months of operation some comment can be made about the issues raised in the hearings and debates.

1. There is evidence that governments in Australia continue to increase the level of censorship of all media, for instance, strengthening the classification criteria (EFA, 1999b, 1999c), the failure to approve a change form X rated videos to Non Violent Erotica and the more restrictive treatment of computer games (OLFC, 1999a).
2. Notwithstanding, the point above the federal government has given up on trying to force ISPs to block overseas content (Deane, 2000). Australian ISPs are now required to offer their customers one of 16 "approved" filter products. The Internet industry appears grateful for this change, and now describe the workings of the Bill as a story of empowerment not censorship (Deane, 2000).

On the other hand others describe the current arrangements as "government endorsed privatised censorship" (Lebihan, 2000; Deane, 2000).

3. It is suggested by the Internet industry that the government "doesn't care any more but we still want to make it work…its good for ISPs to do the minimum required because it will protect us from worse laws" (Deane, 2000).
4. The federal and state "censorship" ministers have conceded that there are difficulties in treating Internet content as analogous to film (Spence, 2000).
5. EFA sought information under Freedom of Information Legislation (FOI) from the ABA to demonstrate the whole system is flawed. After trying for 6 months it received some information much of which was blacked out, and it appeared to consist of many discrepancies (EFA, 2000b). EFA accused the ABA of a "cover-up" (Jackson, 2000).
6. Based on the information obtained under FOI EFA claimed disturbing evidence that value judgements had been made that nudity causes harm but hate or violence against the gay and lesbian community are not harmful. The URLs relating to the latter were not blacked out (EFA, 2000b).
7. For the first three months of operation 124 complaints about Internet content were made to the ABA. Ninety-nine were finalised and 23 complaints proceeding. Two complaints were not investigated as the ABA was of the view they were not made in good faith. Twenty-three take-down notices were issued for domestic hosted content, 45 items hosted overseas were referred to the makers of approved filters and 7 overseas items were referred to the police (ABA, 2000). Inevitably some domestic sites threatened with blockage moved offshore and resumed operations immediately (SMH, 1999)
8. The community advisory panel on Internet content (Net Alert) that was established in November 1999 had not, by July 2000, commenced the public awareness and educational campaign (Hayes, 2000). Much was said about education and empowerment during the hearings and debates especially by the government.
9. Respected people from commerce and industry continue to assert that the Bill does not work and that it damages Australia's international reputation (SMH, 2000).
10. It would appear that the government has now conceded that it is unable to regulate webcasting in the same way as it controls TV broadcasting (SMH, 2000).

REFERENCES

Australian Broadcasting Authority (2000). *Internet Content Complaints Scheme – The First 3 Months*. http://www.aba.gov.au/about/public_relations/newrel_2000/27nr2000.htm date accessed 20 June 2000.

Carol, A. (1994). *Censorship Won't Reduce Crime.* 24 February, 2000, http://www.libertarian.org/LA/censcrim.html.

Commonwealth of Australia (1999). *Broadcasting Services Amendment (Online Services) Act 1999.*

Commonwealth of Australia (1999). *House Hansard: Broadcasting Services Amendment (Online Services) Bill 1999 – Second Reading*, Canberra.

Commonwealth of Australia (1999). *Official Committee Hansard: Senate Select Committee on Information Technologies – Broadcasting Services Amendment (Online Services) Bill 1999*, Canberra.

Commonwealth of Australia (1999). *Senate Hansard: Broadcasting Services Amendment (Online Services) Bill 1999 – Second Reading*, Canberra.

Commonwealth Scientific and Industrial Research Organisation (1998). *Blocking Content on the Internet: A Technical Perspective.* Division of Mathematical and Information Sciences, June.

Cowie, E. (1992). Pornography and Fantasy: Psychoanalytic Perspectives. *Sex Exposed: Sexuality and the Pornography Debate.* L. Segal and M. McIntosh (Eds.), London: Virago Press, pp.132-152.

Deane, J. (2000). *Aussie Censors Zap 27 Sites* ZDNet US, wysiwyg://143/http://www.zdnet.co.uk/news/2000/11/ns-14173.html date accessed 8 September 2000.

Electronic Frontiers Australia (1998). *The Lyons Forum.* http://www.efa.org.au/Issues/Censor/lyons.html date accessed 17 June 2000.

Electronic Frontiers Australia (1999a). *The Net Censorship Dilemma: Blinded by the Smoke - The Hidden Agenda of the Online Services Bill 1999.* http://rene.efa.org.au/liberty/blinded.html date accessed 17 April 1999.

Electronic Frontiers Australia (1999b). *Submission Draft Model State/ Territory Legislation On-line Content Regulation.* http://www.efa.org.au/Publish/agresp9909.html date accessed 11 September 2000.

Electronic Frontiers Australia (1999c). *Community Views – Whose Views? Censorship of Publications in Australia.* http://www.efa/org.au/Analysis/oflcpublrev989.html date accessed 9 September 2000.

Electronic Frontiers Australia (2000a). *The Net Censorship Dilemma: What's New.* http://rene.efa.org.au/liberty/whatsnew.html date accessed 8 September 2000.

Electronic Frontiers Australia (2000b). *FOI Request on ABA Report on Documents Released/ Denied.* http://www.efa.org.au/FOI/efa_foibarep1.html date accessed 9 September 2000.

Felson, R. (1999). Mass Media Effects on Violent Behaviour. *Annual Review of Sociology* 22.

Gibson, P. and Gibson, R. (Eds.) (1993). *Dirty Looks: Women Pornography Power*. London, BFI Publishing.

Graham, I. (2000). *Submission to Senate Legal and Constitutional Legislation Committee: Inquiry into Classification (Publications, Films and Computer Games) Amendment Bill (No. 2)1999*. http://rene.efa.org.au/censor/rdocs/slcsubm0003.html date accessed 5 June 2000.

Hayes, S. (2000). *Net Content Panel Launch*. Australian IT, wysiwyg://139/http://www.australianit...toryPage/0.3811,889354%255E1286,00.html date accessed 8 September 2000.

Jackson, D. (2000). *ABA Accused of Net Cover-up*. Australian IT, wysiwyg://154/http://www.australianit...toryPage/0,3811,743783%255E442,00.html date accessed 8 September 2000.

Lebihan, R. (2000). *Australian Controversy Over Government Web Censorship*. ZDNetUK wysiwyg://142/http://www.zdnet.co.uk/news/2000/26/ns-16352.html date accessed 8 September 2000.

Marr, D. (1999). *The High Price of Heaven*. Allen & Unwin.

Masson, J. (1990). *Incest Pornography and the Problem of Fantasy*. Men Confronting Pornography. M. Kimmel, Ed. New York, Crown Publishers Inc., 142-152.

Morgan, R. (1980). Theory and Practice: Pornography and Rape. *Take Back the Night*. L. Lederer (Ed.), New York: William Morrow.

Office of Film and Literature Classification (1999a). *Guidelines for the Classification of Computer Games: Amendment No. 1.*

Office of Film and Literature Classification (1999b). *Guidelines for the Classification of Films and Videotapes: Amendment No. 2.*

Popper, K. (1969). *The Open Society and Its Enemies: The Spell of Plato*. London: Routledge & Kegan Paul.

Popper, K. (1973). *The Open Society and Its Enemies: The High Tide of Prophecy Hegel and Marx*. London: Routledge & Kegan Paul.

Sandy, G. (2000). The Online Services Bill: Theories and Evidence of Pornographic Harm. *Proceedings of the Second International Conference of the Australian Institute of Computer Ethics*, 11-12 November 2000, Australian National University, Canberra (forthcoming).

Segal, L. (1992). Does Pornography Cause Violence: The Search for Evidence. *Dirty Looks: Woman, Pornography, Power*. P. Gibson and R. Gibson (Eds.) London: British Film Institute.

Spence, J. (2000). *Censorship Ministers Meet*. Ministerial Media Statement 27 July.

Sydney Morning Herald (2000). *Slipping through the Net*. 22 July.

Sydney Morning Herald (2000). *Porn Sites Head Offshore to Beat Law*. 6 September.

Whitaker, R. (1994). *Against the Censorship of Electronic Communication: A Libertarian Argument Against all State Interference in the Provision and Transmission of Pornographic Imagery on Data Networks, Computer Bulletin Board systems and Information Services, and Public Switched Telephone Services.* http://www.libertarian.org/LA/censcomm.html date accessed 24 February 2000.

Whittle, R. (1998). *Executive Summary of the Internet Content Regulation Debate: A Comparison Table between Broadcasting and Internet Communications and a Critique of Peter Webb's Speech.* http://www.ozemail.com.au/~firstpr/contreg/bigpic.htm date accessed 27 March, 2000.

Wilson, E. (1992). Feminist Fundamentalism: The Shifting Politics of Sex and Censorship. *Sex Exposed: Sexuality and the Pornography Debate.* L. Segal and M. McIntosh (Eds.), London: Virago Press.

Chapter 21

The Genetic Revolution: Ethical Implications for the 21st Century

Atefeh Sadri McCampbell, Ph.D.
Florida Institute of Technology, Maryland, USA

Linda Moorhead Clare
Information Technology Group, Maryland, USA

The following chapter defines the practice of DNA analysis and identifies the ethical considerations of human genetic testing in the workplace and technological issues. The topics presented include: the history of DNA testing, future genetic test development inclusive of, large scale, three-dimensional computational technology for analyzing, storing and presenting complex relationships between gene products and clinical outcomes, and currently enacted legislation pertaining to genetic information. Also examined are the ethical concerns pursuant to human genetic testing and information technology security.

A survey was conducted to determine the view and level of knowledge among business professionals in the workplace on the ethical considerations of genetic testing. Questions included business ethics, confidentiality issues and the enactment of state and federal legislation for genetic testing. The results of the survey indicate concern within the business community for ethical issues concerning confidentiality of genetic test results and a strong desire for passage of state and federal legislation to avoid the misuse of genetic information.

Finally, conclusions are determined based upon the data and the direction for future study is defined.

Previously Published in *Challenges of Information Technology Management in the 21st Century*, edited by Mehdi Khosrow-Pour,

INTRODUCTION

According to the Director of the National Human Genome Research Institute, National Institutes of Health, the multi-million dollar effort focused on mapping and sequencing the entire three billion base pairs of the human genome, will be completed in less than a decade (Mansoura et al., 1998). The primary purpose of this massive task undertaken by the Human Genome Project (HGP), is to provide the biomedical research community with the tools necessary to identify the molecular basis of virtually all diseases. Since genetic predisposition plays a role in almost every disease, HGP holds the promise of significant medical benefit for humanity. This hope also raises serious ethical concerns for potential misuses of genetic information. Information technology security is a primary concern because of the threat to confidentiality of genetic test results. According to Ken Sharurette, information security staff adviser for American Family Insurance, "The longer you go without a security breach, the closer you are to your next incident" (Larson, 1999). A number of states have enacted legislation regarding genetic information, but as yet, no fully comprehensive federal laws are in place.

DNA Testing

Genes, the chemical messages of heredity, are working subunits of DNA. Deoxyribonucleic acid or DNA, is a vast chemical information database that carries the complete set of instructions for making all the proteins a cell will need. DNA exists as two long, paired strands spiraled into a double helix. Each strand contains millions of chemical building blocks called bases. There are only four different chemical bases in DNA (adenine, thymine, cytosine, and guanine), but the order in which the bases occur determines the information available, similar to the way specific letters of the alphabet combine to form words and sentences (Klausner et al, 1995). Within the nucleus of every human somatic cell, in two versions distributed over 23 chromosomes (some maternally derived, and the other paternally derived), lies a genetic instruction tape embodied in DNA. Each cell uses this information to create proteins which do the work necessary to carry out the functions of that particular cell type. The sequence is 99.9% identical from one individual to another. This makes it reasonable to speak of a human genome, at least for the purposes of determining a complete representative sequence (Collins, 1997).

Gene testing involves examining a person's DNA, taken from cells in a sample of blood or, occasionally, from other body fluids or tissues, for some anomaly that flags a disease or disorder. Genetic testing in a broader sense includes biochemical tests for the presence or absence of key proteins that signal aberrant genes.

In recent years, many genes involved in hereditary diseases have been identified. The ability to isolate and copy these genes allows biologists to study what goes wrong in cells to cause the diseases. Research on the sequence of all of the

DNA-the genomes-of entire organisms, offers important information on cellular processes and how they are coordinated.

Scientists are working on many techniques to correct faulty genes, including processes to introduce new nucleotide sequences past the body's defense mechanisms. The goal, once the sequences are taken by the cell, is to get them integrated in such a way that the desired substances are properly made. Gene therapy is also beginning to be employed in new and creative experiments that may someday lead to new ways of treating many different disorders. In the future it may be possible to use it to genetically engineer living cells to make their own medicines in response to carefully controlled chemical signals from outside the cell. (NIH Publication No. 97-1051, 1997).

Ethical Implications of Genetic Testing

James Watson, who won the Nobel Prize in Physiology and Medicine in 1962 for codiscovering the structure of DNA, recognized that knowledge derived from genome studies had broad medical and societal implications. As a result, a program was established which was devoted to the ethical, legal and social implications of genome research (Mansoura et al., 1998). The goals of the ELSI program for 1998 to 2003 are:

- Examine issues surrounding the completion of the human DNA sequence and the study of human genetic variation.
- Examine issues raised by the integration of genetic technologies and information into health care and public health activities.
- Examine issues raised by the integration of knowledge about genomics and gene-environment interactions in non-clinical settings.
- Explore how new genetic knowledge may interact with a variety of philosophical, theological, and ethical perspectives.
- Explore how racial, ethnic and socioeconomic factors affect the use, understanding, and interpretation of genetic information; the use of genetic services; and the development of policy. *Text from "New Goals for the U.S.Human Genome Project: 1998-2003," Science 282: 682-689 (1998).*

Issues which are being confronted include the fair use of genetic information; the impact on genetic counseling and medical practice; past uses and misuses of genetic information; and privacy implications of genetic information.. The following high priority areas for research and policy activities have evolved:

- Privacy and Fairness in the Use and Interpretation of Genetic Information.
- Clinical Integration of New Genetic Technologies
- Issues Surrounding Genetics Research
- Public and Professional Education

Consumer groups, in conjunction with the ELSI Working Group, are focusing efforts on the protection necessary to prevent the misuse of genetic information. One initiative calls for enactment of anti-discrimination laws with regard to health insurance and employment. Another initiative focuses on assurance of privacy for persons involved in genetic testing (Mansoura et al, 1998).

Information Technology-Ethical Considerations

In computer systems the weakest link has always been between the machine and humans because this bridge spans a space that begins with the physical and ends with the cognitive. Advanced software and hardware technologies are converging in machine-human interfaces that vastly extend knowledge transfer capacities (Lawton, 1999). There is a rapidly evolving intersection of information, technology and healthcare focused on the integration of information, management of data, representation of knowledge and management of information security.

A major challenge for the entire Human Genome Project is the development of informatic tools to deal with the expected avalanche of project data. The Genome Data Base (GDB) is supported by the National Institutes of Health and is a worldwide repository for genome mapping data that may be utilized as the model for future databases. The main GDB central facility is located in the U.S. and an increasing number of interconnected "nodes" have been established internationally to facilitate access via the Internet by researchers around the world. This is part of the "federated information infrastructure" that allows users to link their computers to a global network of different related databases.

This type of data base development raises ethical concerns regarding security and misuse of information. Even with new, more advanced security available, companies are vulnerable. Internet security systems are focusing on securing data availability, integrity and confidentiality along the length of the entire electronic chain. Global networks, Internet technologies and the demanding pace of change are putting companies and data at risk. Trojan Horses, which mimic familiar programs to trick users' into divulging passwords and other key information, rose in 1999 (Larsen, 1999). It is believed that computer hackers or terrorists are leading causes of security breaches. Should this type of group access a genetic database, the ethical implications are staggering.

Future genetic test development inclusive of, large scale, three-dimensional computational technology for analyzing, storing and presenting complex relationships between gene products and clinical outcomes is underway. The security of this technology should be a high priority for developers.

Technology by itself cannot eliminate risk exposure. It is necessary to have supporting policies, structures and, in the case of genetic testing, legislation in order to orchestrate control over this sensitive data.

Genetic Testing Legislation

A grantee of the Ethical, Legal, Social Implications program and the Department of Energy, developed a model genetic privacy bill which was introduced into the U.S. Senate in November, 1995, and a portion of it has been incorporated into the Genetic Confidentiality and Non-Discrimination Act introduced by Senator Domenici (R-New Mexico) in June 1996. The 104th Congress did not act on this legislation, but a revised version has been introduced to the 105th Congress (Patrinos et al., 1997).

Several states have enacted legislation regarding genetic information and health insurance, but as yet, there are no fully comprehensive federal laws.

President Clinton stated: " We are literally unlocking the mysteries of the human body, finding new and unprecedented ways of discovering not only the propensity for it to break down in certain ways or lead to certain forms of disease or human behavior, but also ways to prevent the worst consequences of our genetic structure. As with every kind of decision like this, there is always the possibility that what we learn can not only be used but can be misused" (White House documents).

Methodology

A total of two hundred questionnaires consisting of nine questions were distributed to business and professional service organizations. The purpose of the survey was to determine the business community's view and knowledge of genetic testing and quantify opinion on the ethical considerations of genetic testing and its role in the workplace. The survey contained questions on the issue of confidentiality of genetic test results for employers, healthcare providers, insurance carriers, and enactment of state and federal legislation with regard to discrimination on the basis of genetic test results. The survey also included demographic questions regarding gender, age, and position within an organization. The respondents were 48% male and 52% female in senior management, human resource management and middle management positions. The demographic focus of the survey was business professionals and the average age was 21-40 years. Analysis is conducted to determine existing and projected trends.

SUMMARY OF FINDINGS

Confidentiality of genetic test results was clearly important to those individuals surveyed. 100% of the respondents stated it would be unethical for businesses or corporations to gain access and view genetic test results before making decisions concerning hiring, job placement and promotions. Also, 96% of respondents did not believe insurance companies should be allowed to use the results of DNA testing to determine insurance approval, coverage and fees. Obviously, the public is concerned that coverage could be denied or fees inflated based upon a genetic test result which would show a predisposition to disease.

Respondents were asked if they believed physicians should have access to DNA test results without consent. The results indicated that 87% of respondents did not believe physicians should have access without consent , while 13% of respondents believed that physicians should have access without prior consent. This statistic portrays a higher confidence level in the ethical behavior of physicians with regard to confidentiality of genetic test results, although a larger number of individuals would prefer to make the decision regarding release of this information to the medical community. This result acknowledges the ethical value of a social practice which physicians generally give precedence to special obligations of confidentiality toward their patients and still concedes that there can be individual cases in which a physicians disclosure of genetic test results for the prevention of harm to a third party would be morally preferable.

Ethical concerns of confidentiality of information is evident in the fact that 93% of survey respondents believe that state and federal legislation should be enacted within the near future prohibiting the release of the results of genetic tests without consent. Some states have already enacted preliminary legislation of this nature, but others have not. Clearly, public opinion is calling for action in this arena.

The survey also asked respondents if they believe state and federal legislation should be enacted prohibiting any person, private firm, private corporation, or private entity from denying or refusing employment to any person or discharge any person from employment based upon results of a genetic test. 87% of respondents responded in favor of enactment of legislation to prevent the denial of employment based upon the results of genetic testing. This question was also asked with regard to the public sector and the statistics are similar, with a result of 89% polled in favor of legislation enacted in the near future.

CONCLUSIONS

As scientific knowledge and technology advance, the collection and storage of human tissues escalates. DNA technology presents society with the promise of significant medical benefit for humanity. Also recognized are the ethical implications of the misuse of genome research and test results. Privacy and fairness in the use and interpretation of genetic information is a high priority for research and policy activities as defined by the Ethical Legal Social Implications Program of the Human Genome Project. The survey results indicate respondents are clearly concerned about privacy of information and non-discrimination in the workplace and commercial sector based upon genetic test results.

Clinicians must endeavor to know what a patient wishes or values are concerning genetic testing. The process of informed consent to genetic testing should include the clinicians disclosure that there may be risk of insurance or employment discrimination.

The Council for Responsible Genetics is a political lobbying group whose goals are the responsible and controlled use of genetic sciences and technology (Macilwain, 1996). The CRG fears governmental abuse or misuse of the one million records contained in the Department of Defense DNA data bank. The lack of substantive procedures regarding military and governmental DNA data banks suggest the possibility of future abuse, such as genetic discrimination inside and outside the military.

It is evident that additional legislation needs to be implemented to counteract potential misuse of genetic information. Also evident is the need for further development of policies, procedures and structures supporting technology development and security of information.

Direction for Future Study

Further research should be conducted to determine the impact of the current legislation regarding genetic test results on the issues of confidentiality and discrimination. The public should also be polled to determine importance of legislative issues and action to be taken.

REFERENCES

Collins, F. (1997). *Sequencing the Human Genome*, Hospital Practice, Vol 32 p.35

Eisenberg, R. (1994). Technology Transfer and the Genome Project: Problems with Patenting Research Tools, *Risk: Health, Safety, and Environment* 5, pp. 163-74.

Fox, J. (1997). *Forget Washington: State Laws Threaten to Restrict Genetic Research, Too*, 9 J. NIH Research 19.

Klausner, R. and Collins, F. (1995). *Understanding Gene Testing* (National Institutes of Health, National Cancer Institute, NIH publication no. 96-3905), pp.1-23.

Larson, A. (1999). Global Security Survey Virus Attack. *Information Week*, Issue 743, pp. 42-56.

Lawton, G. (1999, February). Building the new knowledge interface. *Knowledge Management*, pp. 45-52.

Macilwain, C. (1996). *U.S. Military Tightens Rules on DNA Records*, Nature, Issue 380, p. 570.

Mansoura, M. K. and Collins, F. S.(1998). Medical Implications of the Genetic Revolution, *Journal of Health Care Law & Policy,* 1 (2), pp.329-352.

Meslin, E.M. et al. (1997). The Ethical Legal, and Social Implication Research Program at the National Human Genome Research Institute, Kennedy Institute Ethics, p.291.

Patrinos, A. and Drell, D. (1997). Introducing the Human Genome Project: Its Revelance, Triumphs, and Challenges, *The Judges' Journal*, 36 (3), pp.5-9.

President William Jefferson Clinton, Remarks at the Genetic Screening Event (July 14, 1997). (transcripts available at the "White House Virtual Library" web site (visited August 12, 1999) http://library.whitehouse.gov/ by searching the "White House documents" site using the phrase "genetic" and the date "July 14, 1997").

U.S. Department of Health and Human Services (1997). Public Health Service, NIH, National Institute of General Medical Sciences, NIH Publication No. 97-1051, *Inside the Cell* pp.4-6.

About the Editor

Alireza Salehnia has a Ph.D. in Technology Teaching from the University of Missouri, Columbia, 1989, an M.B.A. in Business Administration from the Central State University of Edmond, Oklahoma, 1977, and a B.A. in Cost Accounting from the Iranian Institute of Advanced Accounting, 1975. He has been a professor of Computer Science at South Dakota State University, Brookings since 1998 and, since 2001, is the acting department head of the Computer Science Department at South Dakota State University, Brookings. From 1984-1989, Dr. Salehnia was an assistant of the Computer Science Department at Culver-Stockton College, Canton, Missouri; from 1989-1992, he was an assistant of the Computer Science Department at South Dakota State University, Brookings, South Dakota; and in 1992, he was an associate professor of the Computer Science Department at South Dakota State University, Brookings. He has been the coordinator of IBM Partner in Education AS/400 Machine since 2000. Dr. Salehnia was a visiting scholar to NASA-EROS Data Center in Sioux Falls, South Dakota in 1995, and he attended a National Science Foundation Workshop on Ethical Issues in Computing in 1996. He has published more than 30 technical papers in journals and conference proceedings, and has presented more than 35 technical papers in national and international conferences. Dr. Salehnia has served as thesis and/or design/research papers advisor on more than 20 Graduate Computer Science or Industrial Management master's students, and has served on the final oral examination of more than 40 graduate students. He has reviewed 23 textbooks for different publishers, and has served as a member of the Editorial Board of the *Annals of Cases in Information Technology and Management in Organizations*. Dr. Salehnia has reviewed more than 30 technical papers for the *Journal of Information Resource Management*, the *ACM's SIGCSE*, *SIGCE*, and the *Information Management Association International Conferences*. He has served as a member of the program committee for IRMA and ISCA Conferences, as a track chair for an IRMA and ISCA Conference, and as a session chair for IRMA, AoM and ISCA Conferences. Dr. Salehnia's areas of interest are MIS, Database, Operating Systems, and Ethical and Social Issues in Information Systems. He is a member of IRMA, ACM and IEEE.

Index